D0883294

Dmitry Yu. Ivanov
Critical Behavior of Non-Ideal Systems

Related Titles

Smirnov, B. M.

Principles of Statistical Physics

Distributions, Structures, Phenomena,
Kinetics of Atomic Systems

2006
ISBN: 978-3-527-40613-5

Schmelzer, J. W. P., Röpke, G.,
Mahnke, R.

**Aggregation Phenomena in
Complex Systems**

Principles and Applications

1999
ISBN: 978-3-527-29354-4

Lesne, A

Renormalization Methods
Critical Phenomena,
Chaos, Fractal Structures

1998
ISBN: 978-0-471-96689-0

Dmitry Yu. Ivanov

Critical Behavior of Non-Ideal Systems

WILEY-VCH

WILEY-VCH Verlag GmbH & Co. KGaA

The Author

Prof. D.Sc. Dmitry Yu. Ivanov
Department of Physics
Baltic State Technical University
St. Petersburg
Russia

Cover
Jürn W. P. Schmelzer: Niagara Falls

All books published by Wiley-VCH are carefully produced. Nevertheless, authors, editors, and publisher do not warrant the information contained in these books, including this book, to be free of errors. Readers are advised to keep in mind that statements, data, illustrations, procedural details or other items may inadvertently be inaccurate.

Library of Congress Card No.: applied for
British Library Cataloguing-in-Publication Data
A catalogue record for this book is available from the British Library.

Bibliographic information published by the Deutsche Nationalbibliothek
Die Deutsche Nationalbibliothek lists this publication in the Deutsche National-bibliografie; detailed bibliographic data are available in the Internet at
<http://dnb.d-nb.de>.

© 2008 WILEY-VCH Verlag GmbH & Co. KGaA, Weinheim

Typesetting Laserwords Private Ltd, Chennai, India
Printing betz-druck GmbH, Darmstadt
Binding Litges & Dopf GmbH, Heppenheim
Printed in the Federal Republic of Germany
Printed on acid-free paper

ISBN: 978-3-527-40658-6

Foreword to the Russian Edition

La connoissance de certains principes
supplée facilement
à la connoissance de certains faits
Claude A. Helvétius

The physics of critical phenomena, which appeared about 180 years ago as a result of Cagniard de la Tour's and Faraday's first experiments, has never left the forefront of the physics of fluids, and then later, of condensed states physics, as a whole. It is, therefore, quite logical that the recently ended 20th century became a century of outstanding achievements for the physics of second-order phase transitions and critical phenomena. Thousands of works have been published concerning critical phenomena in recent years which brilliantly confirm the main results of modern theory. However, despite this significant success, many questions remain still unsolved. About 40 years ago V. Ginzburg, the future Nobel prize winner, put forward a long-range forecast concerning the development and perspectives of physics and astrophysics, in which he included the problem of second-order phase transitions and critical phenomena in his list of the century's main problems. He particularly emphasized on the behavior of systems whose inhomogeneity was caused by the presence of walls, shear, external fields, etc. The century has ended, but problems concerning the behavior of such *nonideal* systems have turned out to be so complicated that even now, after so many years, these problems have not been conclusively resolved.

Another problem not finally solved yet, which existed for long, relates to one of the most characteristic manifestations of the critical state, critical opalescence. As is known, its name reflects the fact that there is a formal resemblance of the "critical" medium with the milky bluish mineral, opal. This resemblance is not accidental, but is due to a common origin, multiple light scattering on optical inhomogeneities. For critical opalescence such inhomogeneities are the large-scale fluctuations, whose behavior governs all the features of critical phenomena. Moreover, it is quite paradoxical that up to now the results of spectral experiments near the critical point are usually interpreted within the Kawasaki theory, which is not going outside the limits of single scattering mode, while the contribution

Critical Behavior of Nonideal Systems. Dmitry Yu. Ivanov
Copyright © 2008 WILEY-VCH Verlag GmbH & Co. KGaA, Weinheim
ISBN: 978-3-527-40658-6

into the scattering spectral half-width of higher orders is usually used only, at best, as a correction. This is not done just "for the sake of it," as the mathematical difficulties of a consistent consideration of even triple scattering are practically insurmountable.

At the beginning of the 1970s, we carried out an extensive precision experiment on pure liquids in Krichevsky's laboratory. In the course of these investigations, for the first time, the values of three critical static indices were found to transform to their classical values. Since that time, my efforts have been directed to attempts to understand both these problems.

We attributed the restoration of classical behavior in the nearest vicinity of the critical point, which was observed during this experiment, to the influence of gravitation. At that time this was perceived, at best, as a much too daring assumption. Such conservatism was quite natural, as up to 1991, these were the only scientific works in the world which registered such a behavior of critical indices under the influence of gravity. It was traditionally accepted that gravitation obstructs carrying out experiments in the vicinity of the critical point and deforms the critical indices in such a way that they stop to be "true."

However, all the facts presented in this book bear witness to the fact that the presence of an additional factor, such as the influence of gravitation or any other "field," does not "disturb" but actually helps new features of critical behavior to be discovered. Moreover, one must look at this from another point of view, considering that critical index values are not distorted by these fields, but are actually obtained in their presence, and therefore are true values, of course, only under these conditions. And then such a words replacement turns from a question of pure semantics into a way of deeper advance into understanding of the nature of critical phenomena on the whole.

This monograph represents the first attempt to systematize the results of analysis of the influence of a wide range of physical fields (gravitational and Coulomb fields, surface forces, shear flow, the effect of boundaries, etc.) near the critical point on different Ising-like systems (pure liquids, binary mixtures, and magnets). These systems, which become nonideal due to the deformation (suppression) of fluctuations by such fields, begin to fall outside the framework of existing ideas and surprisingly demonstrate, in the nearest vicinity of their critical points, not Ising-like but mean-field, classical, behavior. By "nearest vicinity" we have in mind here the vicinity of the critical point where the influence of such fields is noticeable in highly sensitive and accurate experiments.

Recent experimental and theoretical research on critical behavior of various nonideal systems near the critical point, although being quite rare, fully fits into this general picture whose contours were still outlined in our first set of papers on this theme. By the mid-1990s the situation had changed so much that at the 14th conference on thermophysical properties, which was held in Lyon in 1996, this idea was already greeted with a certain understanding and support. Now, many people believe that this is how the critical behavior of systems in real conditions should be. I believe that the day is not far off, as often happens in science, when the situation changes from impossible to "trivial."[1] One of this book's aims is to facilitate the approach of this moment.

As to critical opalescence, in our case this problem arose at studying of critical dynamics, for which light scattering is the most effective research tool. At that time (the early 1980s) it had already become clear that the existing traditional method for the study of critical opalescence and of different multiplicity scattering, which was based on "successive" approximations, single, double, triple, etc., due to insurmountable mathematical difficulties, was of little promise. Therefore, as a first step we decided to model a situation, using as the model a medium with a regulated extinction, water suspensions of monodisperse polystyrene latexes of various sizes and concentrations.

The task of an interpretation of multiple scattering spectra on Brownian particles and critical fluctuations under conditions of high, but more importantly, constantly growing scattering multiplicity which seemed to be almost impossible at first, nevertheless, turned out to be successfully fulfilled. There was a large measure it to thanks to a lot of work and talent of my colleagues, the postgraduates, and co-workers A. Kostko, V. Pavlov, S. Proshkin, and A. Soloviev. I would like to use this occasion to express my sincerest thanks to them.

The book's content was formed between 1975 and 2000. During this period I had the honor and pleasure to discuss, partly or in full, with many people the problems touched upon here. I would like to express my gratitude to them, while being aware that none of them bear any responsibility for eventual shortcomings in the book's contents.

My warmest and cherished memories go to those who helped me so much in a personal and scientific way but who are alas no longer with us. They are Isaac Ruvimovich Krichevsky, Vladimir Konstantinovich Fedianin, and Yuri Ivanovich Shimansky. I learnt the experimental technique from the brilliant experimentalist L. A. Makarevich. I found discussions with V. P. Skripov and E. V. Matizen, who read the manuscript and made numerous useful suggestions, not only instructive but also really fascinating. The comments of N. A. Smirnova, who also read the manuscript, were extremely precise, original, and in many respects surprising for me. A. I. Rusanov always showed particular interest in my work. My understanding of critical phenomena and light scattering was greatly formed by the very constructive discussions with L. Ts. Adzhemyan, Yu. N. Barabanenkov, A. N. Vasil'iev, I. K. Kamilov, V. L. Kuz'min, F. M. Kuni, Yu. K. Tovbin, A. Ja. Khairullina, D. Beysens, R. Tufeu, W. Wagner, and many others. I express my sincere gratitude to all of them.

I would like to sincerely thank the Russian Foundation for Basic Research for supporting the project of the edition of this book.

And, last but not the least, this book would have never been written if the help from my wife, Tatiana Belopolskaya, was not such permanent, effective, and inspiring.

Saint Petersburg   *Dmitry Yu. Ivanov*
September 2002

---

[1] As mentioned by Arthur Schopenhauer: *All truth passes through three stages. First, it is ridiculed. Second, it is violently opposed. Third, it is accepted as self-evident.*

Foreword to the English Edition

The presence of the above given Forewords of the editor and the author to the Russian edition of the book essentially facilitates my task and I need to add only a little.

First of all I am grateful to Wiley-VCH Publishing House for the honor to be among its authors. It is especially pleasant that my book will be among those that would open the third century of the glorious Wiley history.

In the foreword to the English edition, it is necessary to say a few words about the difference of this text from its Russian original. The research carried out during the past 5 years since the publication of the book in Russian has allowed us to expand essentially the first part of the book, devoted to the statics of critical phenomena under conditions of a real experiment. In addition to the second crossover phenomenon in the nearest critical point vicinity, not less fascinating fact has been added, namely, that the first crossover predicted by the theory "far off" from a critical point is not experimentally manifested for pure liquids. For the first time the question concerning correlations between critical indices and amplitudes is investigated.

All these new results (given in Sections 2.2.9 and 2.3.3) were obtained in recent years via the support of the Russian Foundation for Basic Research (grant no. 06-03-33117), and, taking the opportunity, I once again express my gratitude for this support.

The second part of the book exept for sections 7.5 and 7.6 has remained practically unchanged. However, I would draw attention to a problem of the description of multiple light scattering spectra on Brownian particles and on critical fluctuations, which has found both a theoretical and experimental solution here.

I would like to express my sincere appreciation to Professor Jürn W. P. Schmelzer whose numerous efforts allowed us to present the English version of this monograph in its actual form. I am also grateful to him as to the permanent Chairman of the Dubna Research Workshop "Nucleation Theory and Applications" for his constant interest in my investigations, many of which have been presented in Dubna for the first time.

I could not even expect that my long Russian phrases can be translated into English without any preliminary adaptation. Thanks, Alex.

I am not able to find words that describe the selflessness of my wife during the work on the manuscript! Therefore I simply express my love and admiration to her.

Saint Petersburg   *Dmitry Yu. Ivanov*
January 2008

Critical Behavior of Nonideal Systems. Dmitry Yu. Ivanov
Copyright © 2008 WILEY-VCH Verlag GmbH & Co. KGaA, Weinheim
ISBN: 978-3-527-40658-6

Contents

Critical Behavior of Nonideal Systems. Dmitry Yu. Ivanov
Copyright © 2008 WILEY-VCH Verlag GmbH & Co. KGaA, Weinheim
ISBN: 978-3-527-40658-6

Editor's Preface

The physics of second-order phase transitions and critical phenomena was and is an actively developing field of research. Deep insights into the laws governing these phenomena have been reached in the past and awarded by a number of Nobel prizes. However, a comprehensive picture of these phenomena in real systems is not obtained till now and a number of problems remain open.

The author of the present monograph, Dmitry Yu. Ivanov, is intensively involved in the analysis of these phenomena for several decades. Hereby of special interest for him was the investigation of the effect of fields of different types on the laws governing the critical behavior, a topic which was widely ignored so far. As turned out in the analysis, in the nearest vicinity of the critical point, such fields may restrict the increase of the susceptibility of the systems. As a result the behavior in the nearest vicinity of the critical point is transformed from a fluctuation-dominated behavior, where the Ising model results are applicable, to a region where classical, mean-field behavior may be once more re-established.

This idea was first proposed several decades ago by the author, just as a logically noncontradictory suggestion. Today the situation has changed significantly and although the problem of critical behavior of real systems in the presence of different disturbing factors has still not been definitely solved, recent experimental and theoretical research have added quite weighty arguments in favor of this idea. An overview on experimental data in this respect is given in Chapters 1 and 2 including also a brief historical account of the development of research on critical phenomena. Here also another problem of the theory of critical phenomena is formulated and analyzed by the author, as it seems, for the first time: Why simple power law dependences describe the behavior of fluids so well even for states far away from the critical point? In Chapter 3, the question is discussed whether critical behavior is found at singular points or possibly along certain lines, called "pseudospinodal" curves. Chapters 4–7 are devoted to correlation spectroscopy of multiple light scattering and critical opalescence and their application to the understanding of critical phenomena and beyond.

Critical Behavior of Nonideal Systems. Dmitry Yu. Ivanov
Copyright © 2008 WILEY-VCH Verlag GmbH & Co. KGaA, Weinheim
ISBN: 978-3-527-40658-6

So, in brief, in the present monograph the author gives an overview both on the state of affairs of the analysis of critical phenomena in ideal (in the absence of external fields) systems and of the effect of fields on these phenomena and of a new methodology of experimental investigation of such processes. I am sure that the present monograph will be of interest for all colleagues interested and/or engaged in the analysis of these intriguing phenomena.

Rostock (Germany) and Dubna (Russia)
January 2008   *Jürn W. P. Schmelzer*

Introduction

For the physics of second-order phase transitions and critical phenomena, the recently accomplished 20th century was a period of remarkable achievements. The analyses of problems of this part of condensed matter physics were awarded the Nobel Prize more than once. It was first time awarded to Johannes Diderik van der Waals in 1910 and the last to Pierre-Gilles de Gennes in 1991. No less often, scientists who were actively involved in this field of research became laureates thanks to their scientific work in other areas of physics and chemistry. The beginning of the present century has not seen any change in this trend. In 2003, this most prestigious prize was awarded to Abrikosov, Ginzburg, and Leggett "for pioneering contributions to the theory of superconductors and superfluids."

Thousands of papers have been published on this topic yet. The basic results of modern theory have been brilliantly proven. It could be supposed that such achievements should have already led to the development of a comprehensive picture of what happens close to phase transition points. However, despite such significant successes many problems still remain unsolved. This statement is particularly true with respect to nonideal systems whose inhomogeneity is caused by the presence of walls, flows, external fields, etc. [1]. The problems connected with the investigation of such systems turned out to be too intricate.

This monograph attempts to investigate the behavior of various systems close to the critical point, affected by disturbances of different physical nature, such as gravitational and electric fields, surface forces and shear stresses, sample nonideality, the presence of boundaries, etc. These disturbances shall be further given the general name "field," and its nature will be defined more accurately when necessary. The term nonideal relates to a system affected by, at least, one of these fields. We intend to show that the picture of the behavior of such systems close to the critical point is more complex than usually assumed.

In addition to the first crossover, i.e., the transition from mean-field classical behavior to Ising-like behavior, whose position is set by the Ginzburg criterion, in its nearest vicinity a second crossover, the transition in an opposite direction, can be observed [2]. Thus, it appears that in the nearest vicinity of the critical point, paradoxically, the features of the classical van der Waals-like behavior govern the processes again.

Critical Behavior of Nonideal Systems. Dmitry Yu. Ivanov
Copyright © 2008 WILEY-VCH Verlag GmbH & Co. KGaA, Weinheim
ISBN: 978-3-527-40658-6

It is clear that the related experimental and theoretical material, unfortunately, is not too extensive, but even so it could become the "touchstone" which leads to the specification of new peculiarities of critical behavior. Richard Feynman at one of his lectures –with humor which never left him and as ever surprisingly precise –remarked, "There is always the possibility of proving any definite theory wrong; but notice that we can never prove it right. Suppose that you invent a good guess, calculate the consequences, and discover every time that the consequences you have calculated agree with experiment. The theory is then right? No, it is simply not proven wrong" [3]. Even earlier and more definitely on this theme Albert Einstein stated: "Experiment never says 'Yes' to a theory. In the most favorable cases it says 'Maybe', and in the great majority of cases simply 'No'. If an experiment agrees with a theory it means for the latter 'Maybe', and if it does not agree it means 'No' " [4].

In its most general sense, we are dealing here with the limits of applicability of one or another theory. As always, experiment plays the decisive role. For the critical phenomena theory, experiments performed in the nearest vicinity of the critical point could become such an experimentum crucis. However, the problems one has to cope with in experiments carried out near such special points are not less if it is no more than in theory. Theory deals with more or less ideal systems, while experiment with real ones.

This truth would be absolutely banal if it did not relate to the peculiarities of the matter behavior close to phase transition points. Here, with a relative temperature distance of an order $10^{-3}-10^{-6}$ from the critical point large-scale fluctuations start to play a fundamental role. Their growth is accompanied by a continuous increase in the system's susceptibility to various external disturbances (an almost exhaustive list of which can be found in [1]). As a result, any external influence which is arbitrarily weak under ordinary conditions (gradients, impurities, presence of borders, gravitational and/or Coulomb fields, etc.) approaching the critical point eventually may become a decisive factor. Because of this peculiarity, the requirement of the experimental devices is very high with respect to their susceptibility, stability, and accuracy, and even metrological accuracy does not seem to be too high.

All these features make the performance of experiments close to the critical point extremely difficult, not only to implement, but also to interpret the results. This statement is particularly true for the results obtained in the nearest vicinity of the critical point due to a lack of a fully adequate theory. On the other hand, these circumstances make a precise experiment extremely relevant. In this situation the experimenter can use two fundamentally different approaches. The first approach is to make the experimental conditions as close as possible to the ideal one for which there exists a well-developed theory. When formulating a theory, this approach is absolutely necessary and a very important stage in its development. For the theory of critical phenomena this period started more than 50 years ago and is still continuing, even though, in our opinion, the main features of static critical phenomena, in ideal and almost ideal systems, are very well-known. A very clear illustration of such an approach to studying phase transitions carried out under conditions as close to the model as possible is the launching of an appropriate apparatus into cosmic space with satellites in order to get rid of gravity influence. However, taking into account what has been already said, firstly, some other disturbance will be discovered then, caused by some other physical feature and the critical system, sooner or later, will eventually become nonideal. Secondly, such an approach makes it more difficult to advance the understanding of the behavior of real, nonideal systems. It should be noted that the results of experiments carried out under conditions of microgravity, on the whole, bear out this point of view. The second approach consists of attempts of the direct experimental study of real systems, where the influence of disturbing factors is not fully removed, but in the absence of an adequate theory it would be clear what will happen if we reduce it to a minimum. This method is always more complicated than the first one, but for investigating critical phenomena it has historically been shown to be particularly promising.

The main aim, underlying both parts of the present book, consists in investigating the fundamental features of static and dynamic critical phenomena which manifest itself under real experimental conditions in the nearest vicinity of the critical point. On one hand, there is a significant influence of various physical fields, while on the other hand, the developed multiple scattering on growing critical fluctuations has to be accounted for. The book is structured as follows.

In Chapters 1 and 2, the results of a detailed comparative analysis of various experimental data on the behavior of different systems close to the critical point with the presence of one or another disturbing field (gravity, Coulomb and surface forces, shear stresses, boundaries, etc.) are presented. This analysis leads to the seemingly unexpected conclusion that on approaching the critical point there is continuous growth in the system's susceptibility to external influences which eventually leads to a point where fluctuations are first deformed and then completely suppressed by some of such factors. As a result, the system is found to have mean-field, classical behavior with corresponding critical indices. Apart from this, based on our own experimental data we analyzed the existence of universal relations between critical indices and critical amplitudes. The conditions were also formulated so as to fulfill them in the nearest vicinity of the critical point.

The topic of Chapter 3, although not directly related to the main theme of the whole book, is devoted to a question which is by no means secondary. What we are talking about here is whether the critical point is unique as a point, where the singularity of the fullest set of thermophysical properties appears simultaneously, or it is an ordinary representative of a whole family of such points, forming a special line, the pseudospinodal. On its day, the so-called pseudospinodal hypothesis was proposed and intensively exploited which suggested the existence of a certain line, the "pseudospinodal," where all points would have critical properties. This chapter shows that a line on which singularities of isothermal and adiabatic compressibility and isochoric and isobaric heat capacity would simultaneously appear is forbidden by the laws of thermodynamics. Such simultaneous occurrence of divergences can only take place at separate points, as it is found at the critical point, but such behavior is impossible on a line. Therefore, this chapter is extremely important ideologically. Here the meaning of the term "critical point" returns to its original, very nearly lost, sense. This chapter shows that this meaning refers not to one of many points on the thermodynamic surface, but to the unique one which is located simultaneously on the binodal, spinodal, and critical isotherm.

Chapters 4–6, in the second part of the book, are dedicated to correlation spectroscopy of critical opalescence in developed multiple scattering mode. It should be mentioned that this problem has usually been solved by the method of successive approximations: single scattering, double scattering, etc. This chapter shows that the most effective way of solving this problem is our suggestion to start from the other end of this chain, from really high multiplicity of scattering. Chapter 4 presents a review of the development and current state of critical dynamics. The fundamental works, ideas, and results of this comparatively recent and actively developing direction of the theory and practice of critical phenomena research are examined to a certain extent. Chapter 5 presents the results of physical modeling of critical opalescence using disperse systems with regulated extinction, water suspensions of monodisperse polystyrene latexes with practically ideal spherical particles and of different sizes and concentrations. Here the theory of multiple light scattering spectra developed for the first time on the basis of these model experiments on Brownian particles, at the average, adequately described both our own and other experimental data is also outlined. In Chapter 6, we present both the theory of critical opalescence spectra first developed in the approximation of multiple scattering and the results of the experimental investigation of a strongly opalescent binary mixture (aniline–cyclohexane) also first carried out near the critical point by the method of multiple light scattering correlation spectroscopy. The experimental results agree well "far" from the critical point, in the area of single scattering, with the well-known Kawasaki theory. Our theory was confirmed in the area of strongly developed critical opalescence. As a result, the unique opportunity is opened to go beyond existing restrictions of scattering multiplicity and carry out a real optical experiment not only "far" from but also in the immediate vicinity of the critical point.

Chapter 7 is dedicated to one of the central problems of critical dynamics, the singular behavior of the thermal conductivity, the most distinct singular coefficient near the critical point. Here, the main attention is directed to ammonia whose molecules large dipole moment could potentially, as was noted in the study of static critical phenomena in conducting liquids, significantly modify the features of critical behavior and also the kinetic coefficients. However, the set of experiments consisting of the direct determination of thermal conductivity in the wide neighborhood of the critical point, including the nearest vicinity, and extinction coefficient determination using light scattering with the subsequent calculation of critical indices, compressibility amplitude, and correlation radius, demonstrated, most importantly, an identical character of critical behavior of polar and nonpolar pure liquids near the critical point.

The book is completed with an appendix which, although not directly related to the book's main content, nevertheless, promotes the application of its ideas concerning correlation spectroscopy of multiple light scattering to other systems and directions, and demonstrating the universality of the suggested approaches. Here, we look at the principles of diffusion-wave spectroscopy and the application of the technique of photon correlation in its traditional version for opaque and supercritical systems. The appendix describes the method of simultaneously determining particle size and concentration in high extinction disperse systems using dynamic multiple light scattering. The example of sol–gel systems shows the effectiveness of applying correlation spectroscopy for investigating the dynamics of formation and growth of nanostructures.

Part I

The Statics of Critical Phenomena

1
Statics of Critical Phenomena in the Nearest Vicinity of the Critical Point: Experimental Manifestation

1.1
Short History of Critical Phenomena Research

The problem of the analysis of second-order phase transitions and the critical phenomena related to them, despite of nearly two century long history of their existence and thousands of articles already devoted to them, remains actual till now and is one of the key problems of condensed state physics and physics as a whole [1].

G. E. Uhlenbeck (1900–1988),[1] opening the famous Washington conference of 1965 on critical phenomena [5], started his review [6] with the description of the historical experiments (1861–1869) of Thomas Andrews (1813–1885), who carried out the classical works on the continuity of gaseous and liquid states on CO_2 [7–9]. Later, Andrews not only developed the theory of this phenomenon, but he also developed the very important assumption concerning the universality of matter behavior discovered by him, using CO_2 as an example. He also introduced (in 1861) the concepts of the critical point, critical temperature, and critical pressure. In a parallel development, as early as in 1860, Mendeleev [10] anticipated Andrews's conception of the critical temperature of gases by defining the absolute boiling point of a substance as the temperature at which cohesion and heat of vaporization become zero and the liquid changes to vapor, irrespective of pressure and volume. Later, Mendeleev pointed out that the critical temperature of Andrews is the same as the absolute boiling point introduced earlier by him in 1860 [11].

However, long before Andrews and Mendeleev, in 1822, Ch. Cagniard de La Tour [12] described experiments, which he had carried out in order to prove the existence of a certain maximum expansion limit for liquid, *which after this point will turn completely into steam regardless of applied pressure*. Faraday [13] also came to similar conclusions in 1822–1823 based on his experiments on gas liquefaction. It is these first experiments that should be considered as the beginning of the history of critical phenomena.

1) G. Uhlenbeck (1900–1988) together with
   S. Goudsmith (1902–1979) introduced into
   physics the spin concept of the electron (1925).

Critical Behavior of Nonideal Systems. Dmitry Yu. Ivanov
Copyright © 2008 WILEY-VCH Verlag GmbH & Co. KGaA, Weinheim
ISBN: 978-3-527-40658-6

Van der Waals's thesis (1873), which was also devoted to the problem of the continuity of gaseous and liquid states [14, 15] and for which the author was later awarded the Nobel prize in physics (1910), became an outstanding historical milestone. We can agree with Uhlenbeck [6], and say that this work proposes one of the most remarkable equations, which together with Maxwell's (1874) equal areas rule, describes surprisingly qualitatively truly for such a simple equation the whole liquid–gas area in the phase diagram including both stable and metastable states and the critical point. After that it was clear that there do not exist so-called permanent gases (in the sense that they remain always gases and cannot be liquefied), and that the presence of a critical point, as Andrews suggested, is a universal property of matter.

Van der Waals demonstrated his profound understanding of the real role and meaning of the critical point several years later (1880), when he formulated the principle of corresponding states. The essence of this idea consists in using the critical parameters themselves as units of measurement. Employing them, then the resulting equation of state, in such a reduced form, becomes universal. Apart from everything else, this law played an especially important role in the liquefaction of gases, particularly, of helium (H. Kamerlingh Onnes, 1908).

Over the next few years, other phase transitions, which had amazing similarity with critical point liquid–gas behavior, were discovered. So, in 1885–1889, John Hopkinson found that iron, when heated higher than a certain temperature, loses its magnetic properties and just before this transition its susceptibility has a sharp maximum (Hopkinson effect [16]). This phenomenon was thoroughly examined and described in the classical article by Pierre Curie[2] (1895). The temperature of this transition now bears his name. In 1907, Pierre Weiss introduced the concept of an internal field and explained both the existence of the Curie temperature and the spontaneous magnetization occurrence below it [17].

One more phenomenon which demonstrates the existence of a critical temperature is the order–disorder phase transition in binary alloys, which was first suggested (1919) for crystals by Gustav Tammann on the basis of the increase of electrical resistance in Cu_3Au with rising temperature [18]. Later this type of phase transition was confirmed (1925) for Au–Cu and Pd–Cu alloys via X-ray analysis by Johannson and Linde [19]. W. L. Bragg[3] and E. J. Williams in 1934–1935, using an analogy with Weiss's theory, explained it theoretically [20, 21]. Even earlier, in 1928, a similar result was published by Gorsky [23]. So, this theory should be more correctly called the Gorsky–Bragg–Williams theory even more taking into account that, to Bragg and Williams, Gorsky's work was known.

All three classical theories (the van der Waals, Weiss, and Bragg–Williams theories), despite their formal differences, have, as became clear, the same important unique feature: the attracting forces between molecules, which guaranty the

---

2) Pierre Curie (1859–1906): Nobel-Prize laureate for physics (1903) for his research on radiation phenomena (together with M. Sklodowska-Curie (1867–1934)).

3) W. L. Bragg (1890–1971) together with his father W. H. Bragg (1862–1942) are Nobel-Prize laureates in physics (1915) "for their services in the analysis of crystal structure by means of X-rays."

cooperative effect, are assumed to be long range. As a result, these types of theories turned out to be equivalent to the model of interacting particles with an infinite interaction radius (see, e.g., [24]). They bear the general name *mean-field theory*, which was adopted from Weiss's theory, and predict exactly the same singularities for the transition point [25]. As these theories do not predict unstable regions (see, e.g., pp. 127–130 in [26]), the need of the van der Waals isotherms in Maxwell's rule disappears.

Lev D. Landau first pointed out the general connection between these kinds of phase transitions (second-order phase transitions) and the change in the system's symmetry. He developed (1937) a quantitative theory of such transitions,[4] based on the expansion of the thermodynamic potential into a power series of the so-called order parameter, representing the deviation of some thermodynamic quantities in the unsymmetric from those in the symmetric state. The possibility of such an expansion was suggested a priori [25, 27]. The Landau theory, which is the most general formulation of the classical ideas, continues up to now to maintain its role as a universal zeroth-order approximation in the physics of phase transitions and condensed states.

As for experimental researches of critical phenomena, the works of R. Gouy (1892–1893) [28, 29] should be mentioned first. Carried out long ago, these works have much outstriped the experimental level of that time. Gouy was the first to describe, with surprising perspicacity, the influence of the gravitational field on matter close to the critical point (see below for more). A great contribution to the understanding of the critical state was made by A. Stoletov at the end of the 19th century. His critical articles on this problem [30, 31] contain a huge amount of precise observations, remarks, and judgments and are still relevant up to now. A valuable continuation of Gouy's experiments was the extensive series of papers by a group of Canadian researchers (see, in particular, [32–34]). In the mid-1950s, by increasing the level of experimental precision, they methodically studied the thermodynamics of the critical point, carried out spectacular research on the influence of gravity on the coexistence curve in vessels of large and small height [32, 33] (see below for more).

When talking about the nature of critical phenomena we must not forget to discuss the behavior of correlation functions. Correlation function properties take on a particular importance due to three phenomena observed near the critical point, which initially seem disparate, but are in reality closely connected. They are (i) amplification of density fluctuations, which von Smoluchowski proposed (1908) as the reason for critical opalescence [35], (ii) increase of compressibility [14, 15], and (iii) increase of the radius of action of density–density correlation functions (L. S. Ornstein, F. Zernike[5] (1914) [36]). The Ornstein–Zernike (O–Z) theory [36] was the first which could explain the anomalous growth of forward scattering when

---

4) Lev D. Landau (1908–1968): Nobel-Prize laureate in physics in 1962 "for his pioneering theories for condensed matter, especially liquid helium."

5) F. Zernike (1888–1966): Nobel-Prize laureate in physics (1953) for his demonstration of the phase contrast method and invention of the phase contrast microscope.

investigating critical opalescence (see Chapters 4–6 for more) and paid attention to the presence of a significant long-range part in the correlation function. However, it is this circumstance which makes the O–Z theory classical, in the above-mentioned sense, as it is also based on the suggestion concerning long-range forces, and fluctuations are only considered as small corrections.

The first nonclassical result, which was practically the only one at the time of the Washington conference, was obtained by Lars Onsager[6] (1944) [37] in the analysis of the two-dimensional Lenz–Ising model, the so-called nearest-neighbor model with clearly short range forces (the history of the Lenz–Ising model can be found in the article [38]). Despite the fact that Onsager's most impressive result was the prediction of the logarithmic singularity of isochoric heat capacity in the neighborhood of the phase transition point instead of a classical "jump," this result was considered in the 1940s–1950s to be a mathematical curiosity rather than a physical reality. Such an approach to it continued until, in the 1960s, such type of behavior was discovered for the heat capacity of helium, argon, and oxygen (see, e.g., [39, 40]). So, in the early 1960s the appearance of additional experimental proofs of the inadequacy of the critical phenomena description by classical theories not only stimulated a renewal in interest and further rapid growth in their development (this explosive growth was noticeable, despite that even thousands of investigations into critical phenomena in liquids carried out between 1950 and 1967 (see, e.g., [39])) but also marked the beginning of the new modern stage in the research of phase transitions.

In general, as correctly mentioned by many authors (see, e.g., [26,41]), it is difficult to fix any certain date by which the modern period in studying critical phenomena has begun. Such a point could possibly be either the Onsager's exceptionally important work [37], if we agree with E. Stanley [26], or E. Guggenheim's famous analysis [37] which demonstrated a cubic rather than a square shape of the coexistence curve for eight simple liquids. Or maybe, as M. Fisher said [41], it was slightly earlier when the same result was obtained in another well-known research for CO_2 [43].

Undoubtedly, this would all be true if it was not for one extremely important circumstance. In fact, in the 1890s van der Waals found that experimental data on the critical behavior of surface tension showed an unusual (what we would now call "nonclassical") value of the critical index. Actually, the very idea of critical indices was also not new, but was introduced by van der Waals himself (1893) for describing just the critical behavior of surface tension. Van der Waals was not the only one in detecting deviations from his own theory. We must not forget the Belgian researcher J. E. Verschaffelt who, as a young man, came to the Netherlands on a 2-year apprenticeship in 1893, began working with J. van't Hoff, the first Nobel-Prize laureate for chemistry (1901). He then continued working with van der Waals

---

6) L. Onsager (1903–1976): Nobel-Prize laureate in chemistry (1968) for the discovery of the reciprocity relations bearing his name.

and, finally, ended up at the laboratory of H. Kamerlingh Onnes,[7] which was one of the best laboratories of the time. He stayed there for many years. As a result of applying an analysis technique, the logarithmic differentiation method, developed by him [45] on his own data as well as on already published experimental data, Verschaffelt in 1900 (!) as obtained practically modern critical index values for pure liquids. (It is interesting that the logarithmic differentiation method was again rediscovered in the 1960s and it is called now the Kouvel–Fisher method [44]). Critical index values obtained by Verschaffelt: $\beta \sim 0.34$ for the coexistence curve, $\delta \sim 4.26$ for the critical isotherm, and $\bar{\mu} \sim 1.32$ for the surface tension coefficient, instead of their classical values 0.5, 3.0, and 1.5, respectively. Besides, Verschaffelt managed to develop an equation of state similar to modern equations of state close to the critical point. But, as J. M. H. Levelt Sengers wrote in [46], from where all this information about Verschaffelt's work is taken, all this, unfortunately, fell "on deaf ears." Looking at the reasons why Verschaffelt's work was largely ignored Levelt Sengers comes to the seemingly correct conclusion that this was "the seed on the rock." Every idea needs to be mature and in demand. It is also suggested that the fact that practically all of Verschaffelt's works were written in Dutch played a negative role. Levelt Sengers, who did a great work for the physics of critical phenomena, took over the task of translating and commenting on these works. This was a difficult but useful task [46] which typically also passed almost unnoticed.

So, we can see that the reasons for looking again at the classical theory of critical phenomena arose almost at the same time as its creation, but fate disposed differently. It was necessary to wait for a really not ordinary event, the appearance of L. Onsager's work [37], for the reasons of inadequacy of all classical theories of phase transitions to became clear. This created the certainty that it was possible to understand them deeper and pointed to the directions which should be followed. As C. Domb and M. S. Green, the first editors of the famous multivolume collection "Phase Transitions and Critical Phenomena" wrote in the preface: 'The problems to be faced in extending Onsager's work were formidable, and at first progress was slow. However, there were many different aspects to be investigated and a number of alternative lines of theoretical approach to be pursued. The field began to attract new experimentalists and theoreticians of ability' [47].

Finally, it was this last circumstance that was the deciding factor. A new modern stage in the development of views on the nature of phase transitions and critical phenomena, in which not the classical but the fluctuating approach to describing them dominates, began to develop and still continues today. The essence and details of this approach are presented in depth and fully in the reviews [39, 48, 49], in the books [26, 40, 41, 50–56], and also in the above-mentioned encyclopedic series [47]. Therefore, below, within the limits of our brief historical review, only main ideas underlying modern representations concerning the nature of phase transitions and the critical phenomena will be noted. These ideas [57–64] were formulated almost

---

7) Heike Kamerlingh Onnes (1853–1926): Nobel-Prize laureate in physics (1913) "for his investigations on the properties of matter at low temperatures which led, inter alia, to the production of liquid helium."

immediately after the 1965 Washington conference [5], where their appearance was so wisely predicted by Uhlenbeck (see below). Moreover, it was in those years that significant experimental material [26, 39–41, 48, 49, 65, 66] was obtained, which did not play a small role in the establishment, development, and enrichment of the theory.

By "fundamental ideas" we are first of all thinking of the scaling hypothesis (or *scaling*), which makes it possible, in particular, to understand the fact of the universality of second-order phase transitions and critical phenomena. This idea was formulated almost simultaneously and independently by B. Widom (1965) [58] for liquid–vapor systems, by C. Domb and D. Hunter (1966) [59] for the Ising model, by A. Patashinsky and V. Pokrovsky (1966) [60] for ferromagnetic and in an apparently clearer physical form leading to an understanding of the universality of critical phenomena (see, e.g., [51, 67]) by L. Kadanoff (1966) [61] for correlation functions.

Once more, all this would correspond to reality if it were not for one fact. Today, when the idea of the universality and generality of second-order phase transitions and critical phenomena has become generally accepted and is beyond any doubt, we must remember V. Semenchenko whose fundamental ideas were always based on the general fluctuating nature of these phenomena. As a result, he was the first, in 1947, to formulate the idea of their thermodynamic generality [68]. However, as one of his brightest pupils and followers, V. P. Skripov, wrote: '. . . at the end of the 40's and 50's the idea came up against clear or hidden opposition by physical chemists. The seed fell on stony ground' [69]. So, unfortunately, the author of the idea to unite critical phenomena and second-order phase transitions based on fluctuation theory concepts suffered almost the same fate as Verschaffelt.

The idea of the universality of these phenomena is inherent in V. Ginzburg's[8] already classical work [70], where different second-order phase transitions which have a clear heat capacity anomaly (liquid helium, alloys, etc.) and those which have not (superconductors, ferroelectrics, etc.) are considered as the same-type transitions, transitions of the same physical nature whose difference is of an exclusively quantitative character. Ginzburg succeeded in taking this idea to its logical final and proposed the criterion which now bears his name and which, in principle, makes it possible to predict when the "normal" character of the transition is switched to fluctuation determined [70].

At the beginning of the 1970s, Kenneth Wilson made the next step. Starting his analysis with Kadanoff's "blocking picture" [61] and using then the ideas of the renormalization group method (RG) developed in the relativistic field theory in the 1950s, Wilson managed to transform the semiphenomenological ideas of scaling and universality into a real calculation scheme close to the critical point [71]. Then F. J. Wegner showed [63] how universality and scaling naturally followed from Wilson's method [72]. A year later, Wilson and M. E. Fisher succeeded in completing a self-consistent calculation method within the limits of RG by

---

8) V. L. Ginzburg (1916): Nobel-Prize laureate in physics in 2003 "for pioneering contributions to the theory of superconductors and superfluids" together with Alexey A. Abrikosov and Anthony J. Leggett.

introducing the so-called ε-expansion [73]. They showed that if one introduces variable space dimensions $d = 4 - \varepsilon$, then it becomes possible to calculate indices in the form of a series of ε. Therefore, in order to describe critical behavior there was created "a new language no less universal than Landau's old theory, but more flexible and possessing a great 'predictive power' " [51]. In 1982, Kenneth G. Wilson was awarded the Nobel Prize "for his theory for critical phenomena in connection with phase transitions."

In no way negating K. Wilson's merits, L. Kadanoff should, unarguably, also be recognized for completing a possibly more important, though intuitive, advance in the route to working out critical phenomena theory. He postulated universality classes and scaling hypothesis, based mainly on the fact that the only defining length in the critical region is the correlation length and also on the idea that the symmetry of interactions, but not their details, determines the critical behavior. The RG-theory successfully used these ideas, thereby confirming the validity of Kadanoff's foresight. Later the RG-theory was developed thanks to the works of many people and it is now an extremely powerful instrument which makes it possible to calculate many different properties: critical indices, critical amplitude ratios, universal correlation functions, equations of state, etc. (see, e.g., [50–55, 67, 74–78]).

These theoretical successes were closely related, as already mentioned above, to the progress made in experimental researches close to the critical point: the pioneering works of Skripov and Semenchenko [79], Voronel and Amirkhanov with their colleagues on heat capacity [80, 81], Skripov with his colleagues on the use and study of light scattering (in the prelaser era) [82], Shimansky and his colleagues on the wide research of the static properties of matter [83], Makarevich from the Krichevsky laboratory who brought a technique of (pVT)-experiments to perfection [84], Krichevsky's thermodynamic works on critical phenomena in dilute solutions [85], which have outstripped time much more (see, e.g., [86]). Ivanov et al.'s precision experiments [87, 88] could also be added to this list, where for the first time coexistence curve asymmetry and change in the static critical indices in the nearest vicinity of the critical point to the direction of their classical values were attributed to gravitation [89–92].

Daniel Beysens and his colleagues in Saclay (France) carried out the encyclopedic work on investigating the features of static critical phenomena using the example of binary mixtures (see, e.g., [93, 94]). Beysens also initiated and carried out the first experiments on research on the critical behavior of binary mixtures in flow [95], which gave, in principle, the same result as the experiment on pure liquids with gravitation [87, 89], transformation of critical indices to their mean-field values (see also [96, 97]). At the same time the fundamental theoretical Onuki–Kawasaki paper [78] appeared. In this work, within the framework of the RG-approach, the authors showed that a system influenced by shear flow crossed over to mean-field behavior on approaching the critical point due to deformation, up to suppression, of the long-wave critical fluctuations. As a consequence, critical indices get classical values.

Finally, we should mention the research of Wolfgang Wagner's group [98, 99] in which, as it earlier was in our studies [87, 89, 91], during the brilliantly organized, fully automated $(p\rho T)$-experiment on SF_6 [98] in the nearest neighborhood of the critical point the classical values for three (β, γ, δ) static critical indices were obtained. The reason put forward for this behavior was gravitation. Subsequently, by repeating their experiment on CO_2, they not only showed that SF_6 is not an exception but also that the change of indices to their classical values can be directly attributed to gravitation [99].

Referring for the last time to the Washington conference (1965), which can be considered to have concluded one classical, era of critical phenomena, and actually initiated the new era related to the fluctuation theory of phase transitions, we quote again the introductory remarks by G. Uhlenbeck [6]: "If there is such a universal, but nonclassical behavior (*of heat capacity singularity*, D.I.), then there must be a universal explanation which means that it should be largely independent of the nature of the forces. The only corner where this can come from is I think the fact that the forces are not long-range ... With regard to the correlation function and the Ornstein–Zernike theory I hope that the new experimental work will show that there is a *critical region*, where already deviation from the classical van der Waals–Ornstein–Zernike theories begin to show up. I think there are indications (as the specific heat anomaly) that such a region exists." Let us add another final remark by Uhlenbeck concerning nonequilibrium phenomena: "There are very surprising experimental results found about various transport coefficients,[9] which clearly are worth pursuing. Till now there is no theory, not even a van der Waals like theory" [6].

It really was another era, although only a few decades ago. It was also different, as today after Uhlenbeck's prediction has come brilliantly verified, this critical area was not only discovered but has also been analyzed in detail. We hope to show that (see Chapters 1 and 2) beyond the critical area, even closer to the critical point, classical behavior returns once more! Maybe, in the grand scheme of things, Krichevsky, who was the most thorough and consistent supporter of the classical views on critical phenomena and kept his faith until his death, was not so wrong. Maybe he was just as wise as Newton with his "corpuscles." Fortunately, the history in both cases developed in their own way without terminating the appearance and development of other ideas. As a result we have a perfect wave theory of light and also the timely appeared quantum theory. Finally, quantum electrodynamics could once more explain "everything" using only "corpuscles" [100]. The theory of critical phenomena has not yet reached such perfection but we hope that it is close.

This short historical excursion[10] is, undoubtedly, necessarily very subjective, but the presence of a great number of excellent reviews and books on this theme [39–41, 46, 48–53, 55, 56, 65, 66] significantly simplified the author's task and gave the opportunity to look at the key moments and little-known facts from the history of the development of ideas concerning the nature of critical phenomena. Anticipating the passage to the description of the basic theme, it should be noted that, unlike most well-known monographs and reviews (theoretical and experimental) whose

---

9) For more details on the behavior of the transport coefficients close to the critical point, see Chapters 4 and 7.

10) It should also be noted that Chapter 4 is concentrated on a review of research and ideas related to the dynamics of critical phenomena.

authors did not look at all at the influence of external factors on the behavior of critical phenomena or interpreted them as an obstacle in the way of investigating a "true" critical behavior, the first two chapters, and most of the whole book, are dedicated to the evolution of critical behavior of various nonideal systems as they approach the critical point under an influence of fields of different physical nature.

1.2
Peculiarities of the Experiment in the Nearest Vicinity of the Critical Point

1.2.1
"Experimental" Critical Indices

In the asymptotic vicinity of the critical point, independent of whether a classical or fluctuation type description is chosen, different physical quantities can be represented as simple power-law dependences. These exponents, critical indices introduced already by van der Waals, play a central role in any critical phenomena theory, as one or another set of their concrete values determines one or another type of the critical behavior. An exact critical index determination looks like this [26, 101]:

$$\lim_{x \to 0+} \frac{\ln f(x)}{\ln x} = \lambda \quad \Rightarrow \quad \underset{x \to 0+}{f(x) \propto x^{\lambda}}. \tag{1.1}$$

At the beginning of the 1970s, it became clear on the basis of experiments on liquids and the RG-analysis [63, 102–104] that precise experimental data cannot be described by simple power-law dependences and in general $f(x)$, which represents one or another physical (thermodynamic) quantity in the wide vicinity of the critical point, should be represented in a more complex form as

$$f_i(x) = \Lambda_{0i} x^{-\lambda_i} \left[1 + a_1 x^{\Delta} + O\left(x^{2\Delta}\right) + \cdots \right] s, \tag{1.2}$$

where x is the relative distance from the critical point in the thermodynamic variables, λ and Λ_0 are critical index and amplitude, while Δ and a_1 are the exponent and coefficient for the correction term, respectively.

Critical indices, unlike the full functions like Eq. (1.2), are experimentally measurable, which is one additional reason why they are playing such an important role. Really, as the polynomial behavior (Eq. (1.2)) close to the critical point is mainly determined by its leading term, so in accordance with definition (1.1), by the slope in the log–log scale graph of the dependence $f(x)$ on x, the critical index λ can be determined. Thus, the obtained critical index can generally be considered as an *effective*. Two circumstances should be mentioned in this connection. First of all, it can be easily shown (see, e.g., [104]) that the real (λ) and effective (λ_{eff}) critical indices are connected by a rather simple relation (see the very interesting papers [105–107] devoted to this subject):

$$\lambda_{\text{eff}} \cong -\frac{\partial \ln f}{\partial \ln |x|} \cong \lambda - a_1 \Delta \, |\overline{x}|^{\Delta} + O\left(|\overline{x}|\right)^{2\Delta} . \tag{1.3}$$

Here $x = \tau$, $\tau \equiv (T/T_c) - 1$, where T_c is the critical temperature, and $\bar{x}$ is some average temperature of experiment. Taking into account the fact that $\Delta \approx 0.5$ [102, 103] and $a_1 \leq 1$, then for the mean value $\tau \sim 10^{-4}$ the difference between effective and "true" critical indices does not exceed 0.005, which is well beyond the measurement error. Secondly, as was shown in [108], effective critical indices in leading order in ε obey all scaling relations (apart from hyperscaling one, which includes a space dimension). Below, we shall show, in the course of discussion of the precision of (pVT)-experiments on pure sulfur hexafluoride, that this point of view, which is only justified for ideal experimental conditions, should be corrected for real systems. Another no less important reason why critical indices receive such attention is related to the fact that thermodynamic and statistical fundamentals lead to the existence of a defined set of relations between them, usually inequalities (see, e.g., [26, 40, 41]).

Although thermodynamic conclusions are always categorical, that is why these dependences have a universal and valid character for any particular system, the nonstringency[11] of these inequalities, however, permits them, within the limits of scaling, to pass to equalities without violating any thermodynamic limitations.

Finally, as every model which can be used for describing critical behavior has its own set of critical indices [26, 40, 41, 52], so their experimentally determined values can make it possible to decide which model is preferable. In addition, the theoretical values of critical indices for systems with different values of space dimension (d) and number of components of order parameter (n) are known only as a result calculated for concrete idealized models (like the Ising model, Heisenberg model and their modifications) [26, 40, 41, 48, 52−55, 71−74, 102, 103]. At the same time, the conditions for a real experiment automatically include significant additional factors, especially in the nearest vicinity of the critical point, such as the existence of gravitation, dipole magnetic and Coulomb forces, surface tension forces, etc., which are commonly not considered when calculating critical indices. This is the reason why the comparison of experimental and theoretical values, where the influence of these factors can be very strong and often lead to unexpected consequences, should be made very carefully. It is also important to take into account the experiment's precision, as well as the way of approach to the critical point.

The fluctuation theory of phase transitions, in particular, leads to the conclusion that critical indices only depend on space dimension and the number of components of the order parameter. Transitions with the same number of order parameters form the same universality class [40, 62]. Liquids, solutions, binary alloys, and anisotropic ferro- and antiferromagnets together with the Ising model ($n = 1$) form the same universality class. This is the type of objects that we are examining in this book and in particular we shall discuss their behavior using the example of the liquid–gas critical point.

11) So as to avoid terminological confusion, it should be said that by "nonrigorous" we are referring to the sign "$\geq$." This is often (see, e.g., [26, 40]) called "rigorous" keeping in mind the thermodynamic rigor of their proof.

1.2.2
Determination of Critical Parameters

To process the results of any experiment in the vicinity of the critical point by equations like Eqs. (1.1) and (1.2) it is necessary to know the critical parameters of the substance being used for the experiment. To our opinion, it is not only desirable but absolutely necessary to determine these parameters using the same experimental setup which is being used for the main experiment.

It is sufficient to say that, within the framework of the International Practical Temperature Scale (see, e.g., p. 81 in [109]) it is not possible to determine either the temperature of absolute zero or temperature of reference points better than one hundredth of a kelvin, whereas in the best experiments thermostatting temperature fluctuates only within the range 10 to 200 μK. Therefore, near the critical point it is necessary to use the thermometer's (usually a platinum resistance thermometer) own scale.[12] Moreover, one must determine the distance from the critical point on temperature with one's own thermometer. However, this cannot be accomplished unless the critical temperature is determined during the experiment itself. It is actually quite usual to leave the critical temperature as a free parameter when approximating experimental data by Eqs. (1.1) and (1.2). However, it seems that such a method is forced upon but not effective, and therefore does not always help to obtain an objective picture.

Another peculiarity of experiments in the nearest vicinity of the critical point is that they are carried out against the background of ever increasing system susceptibility to different external factors. That is why there are such high requirements for purity of the investigated substance, for the absence of gradients etc. (see, e.g., [40]). Moreover, it is necessary to have criteria for reaching the critical state, and in order to determine the critical indices needed for comparing the obtained results with theoretical ones it is necessary to obtain values of no less than three critical indices on the same experimental setup. This latter requirement is due to the fact that, although relations existing between critical indices make it possible to consider any two of them as independent, however, the test of these relations themselves demands independent determination of, at least, three indices. To obtain a self-consistent set of indices it is absolutely essential that this requirement is fulfilled.

Further we shall discuss all of these questions and, despite the abundance of works on this problem, the number of research papers which satisfy all the above criteria are not that great. Therefore, we have to use our own experiments carried out on pure SF$_6$ [87–92] and use other authors' papers when possible. As some additional justification of this approach we can say that these researches,

12) It should be noted that when investigating critical phenomena where all these values are referred to the critical point, this condition plays no role. However, it seems that if the investigated samples are properly cleaned, then their absolute value T_c corresponds to each other properly (see, e.g., [98], where the found T_c differs from that determined in [87] by only 1.5 mK, which lies within the margins of error even for absolute temperature measurements).

which began in 1974, were for many years the only ones in the world scientific literature where in the same experiment the changes of three static critical indices toward their classical values in the nearest vicinity of the critical point were found. Moreover, these changes were, for the first time, interpreted as the consequence of the gravitational effect.

For many years this hypothesis was considered, at best, as an overbold one. Many people, including such a great authority on critical phenomena as I. R. Krichevsky, in whose laboratory this experiment was carried out, tended to consider such behavior as a natural consequence of the validity of the classical critical phenomena theory. It was not that long ago that papers appeared where analogous results were once more obtained on SF_6 [98] and CO_2 [99]. In these papers, the authors also treated the transition of the indices to classical values as a consequence of gravitation. The results of these works will be discussed in detail further, but we should start with a brief look at the peculiarities of critical parameter measurement methods and methods of assessing purity of matter.

1.2.3
Purity of Matter

Already in the 1870s, Andrews [7–9] noted that the critical temperature of CO_2 mixed with air can go right down to $0\,°C$ depending on the degree of impurity. In fact, this is so obvious that the only surprise is the reaction against Andrews by such famous scientists as J. Jamin and L. Cailletet who supposed that "the addition of a gas harder to liquefy (air) should not effect the critical temperature of a gas easier to liquefy like carbon dioxide (CO_2)" (see [31], pp. 290–297).

The effect of impurities on critical parameters can be observed by studying diluted solutions. In his thesis, Makarevich [110] determined the critical parameters of diluted solutions of carbon dioxide in sulfur hexafluoride. By analyzing his results we managed [141] to ascertain that a good test on matter purity can be the so-called *critical coefficient*, $K = (RT_c/p_c V_c)$, where the subscript "c" denotes "critical," R the universal gas constant, $R = 8.3145\,\mathrm{J\,mol^{-1}\,K^{-1}}$ (Fig. 1.1). The figure clearly shows that as the amount of CO_2 decreases this coefficient increases. As this coefficient continues to grow as other impurities are removed, it can be supposed that this feature relates not only to SF_6 and its impurities, but has a more general character. Unfortunately, we are not aware of such tests carried out on other substances.

As for the two purest of the tested samples (matter purity 99.9994% [98] and 99.9995% [87–92]), although only semiquantitative evaluations were made, the dependence found in [141] turned out to be very good (Fig. 1.1). Clearly, these conclusions are valid only if the proposed criterion continues to operate even at such a low degree of impurity and if there were no systematic errors in determining the critical parameters in the quoted papers. It is in any case clear that in [91,98] we are dealing with the purest ever investigated samples of sulfur hexafluoride (and not only of this).

In addition, the evaluation of the degree of purity performed in [87,91], as for [110] in its time, was carried out using one of the typical features of pure matter,

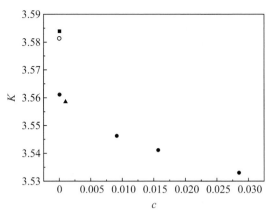

Figure 1.1 Dependence of the critical coefficient of SF_6 on impurity concentration [141]. Plots are made using experimental data from different sources: (•) SF_6 and CO_2 [110]; (▲) data from [113] (quoted from [114]); (■) [87, 91]; (○) [98].

the horizontality of its isotherms in the coordinates $(p-\rho)$ in the two-phase region. After a double purification by the first rectification column the pressure difference at the ends of the isotherm with a four times change in volume did not exceed ± 0.5 kPa. The purification by the second column stopped only when the pressure difference, under the same conditions, was not noticeable within the limits of the sensitivity of the pressure gauge (± 2 Pa). To do this it was enough, as a rule, to carry out two to three distillation cycles on the rectification column for a final purification [91].

With precision thermostatting and a sensitively stable pressure gauge this test, although not providing a quantitative or qualitative estimate of the composition of the impurities, has, in our opinion, an incontestable advantage over other methods (chromatography, mass spectrometry, etc.). Most importantly this check is carried out on the same experimental setup which is then used for the main experiment. This makes it possible to carry out a closed cycle, which guarantees that impurities will not enter the checking process itself and also its efficiency. Finally, this method is universal, while quantitative analysis methods are to different degrees developed for different substances. In a recently published work [111], one of the authors of which is the renowned researcher in the field of critical phenomena, L. Weber (see for example one of his early works [112]), not only such method of definition of substance purity is proposed, but also detailed recommendations for its calculation are given.

1.2.4
Determination of Critical Density

One of the most difficult problems of determining critical parameters is critical density, (ρ_c). This is due to the natural peculiarities of the critical point. The small curvature of the coexistence curve close to the critical point leads to low pressure sensitivity to density changes. It is also due to terrestrial observation conditions when the gravitational effect [28] causes additional flattening of the peak of this curve [32–34, 88, 89, 91, 115]. All this leads to the fact that, even in the best works, the precision of critical density determination is, as a rule, two to three times lower than the determination of other critical parameters and does not exceed $\pm(0.1–0.2)\%$. In our research the precision of critical density determination was of an order higher, $\pm0.02\%$ (see below for more).

The most common method of determining critical density is the disappearance of the meniscus method [28, 31–34, 84, 116, 117] or the so-called Cailletet and Mathias "rectilinear diameter" rule (1886) [118]. In recent years, however, the validity of this rule for the nearest vicinity of the critical point has raised such serious doubts that the diameter of the coexistence curve was even given its own name "singular" (see, e.g., [119–124] and also [125] and references therein). Therefore, it was necessary to have a method for independently determining the critical density. Such a method, in view of all previous experience (see, in particular, [31]), was suggested by Makarevich and Sokolova [84] and then used in [87–92]. In view of this problem's great importance it is necessary, albeit briefly, to look at the principal peculiarities of this method of determining critical density.

The method is based on using visual observation of meniscus movement in a constant-variable volume piezometer. The high precision of critical density determination ($\pm0.02\%$) in this case is due to both the high sensitivity of such a method of observation, as Gouy [28] and Stoletov [31] still noted, and the improved sensitivity of the volume measurements as a whole [84]. The sharp increase in sensitivity and precision of the volume measurements in its turn was possible only thanks to the making and application of the micropress, an original refinement of the (pVT)-setup. The micropress is a successful technical realization of I. R. Krichevsky's fruitful idea which makes it possible to unite the advantages of constant and variable volume piezometers in one apparatus [84].

Let us assume that the critical parameter values of the substance being investigated have been roughly determined (this is not usually very difficult). Then, an appropriate amount of the substance is placed into the piezometer and the temperature of the thermostat is raised by $(3 - 5) \times 10^{-3}$ K and is further kept at this level with extreme precision (in this case $\pm2 \times 10^{-4}$ K). Next, using the micropress the volume of the piezometer is quickly (quasiadiabatically) increased by $(0.3–0.5)\%$ (AB, see Fig. 1.2) from the whole volume which leads to cooling of the substance (BC), appearance of the second phase, and the meniscus.

Then, the volume is slowly decreased down to the initial one (CD), at the same time generating an undulatory motion of the meniscus by light shocks of the stirrer. The less intensive this motion, the greater is the time for transition to equilibrium. In particular, the motionless meniscus is dissipated in at least 2 h, while with intensive stirring it disappears within several seconds against the background of very strong opalescence. The speed of the meniscus undulation

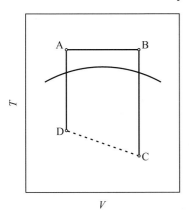

Figure 1.2 Schema of determination of the critical density by the disappearance of the meniscus method.

increase was experimentally chosen in such a way that in 1–2 min the meniscus begins to darken, finally turning into a 1-mm-thick dark brown stripe with very weak opalescence in the rest of the vessel. These observations demonstrate that without stirring the critical phase inside a piezometer can, apparently, exist only in a thin layer of the substance.

The transition from a nonequilibrium state (D) to an equilibrium one (A) is accompanied by the meniscus moving up or down the vessel's height depending on the value of the mean density of the substance therein. When $\overline{\rho} > \rho_c$ holds, the meniscus moves up and when $\overline{\rho} < \rho_c$ is fulfilled, it moves down. Thus, the task is reduced to find the position at which the meniscus movement changes to the opposite at the minimal change of average density (volume). The sensitivity of such a method is limited only by the micropress's sensitivity which in [84] and then in [87–92] was not worse than 0.002%. If we take into account the limits of error of the micropress's and piezometer's calibrations, then the error in determining the critical density did not exceed ±0.02%. It should be mentioned that the possible systematic influence of gravitation on the result of the measurements, naturally, is not included in this estimation.

1.2.5
Determination of Critical Temperature and Pressure

The precision in determining the critical temperature (T_c), using the same visual observation method for the appearance–disappearance of the substance's two-phase state in the piezometer when the average density being equal to the critical one, was not worse than ±2 × 10^{-4} K with thermometer sensitivity no less than 2 × 10^{-5} K [87, 91]. The adjusted values are, of course, not related to the absolute scale but only to the own one of a secondary reference platinum resistance (100 Ω) thermometer. Wagner and his colleagues did not carry out visual observations, but temperature instability was reduced to 10–25 μK [98, 99]. The measuring system of pressure, including critical one (p_c) used in our experiment [87–91], consisted of a piston gauge, belonging to the group of the national standards, a mercury null indicator, and a reference control mercury barometer (for measuring changes in atmospheric pressure during the experiment). This guaranteed an overall error not exceeding ± 40 Pa ($\sim 0.001\%$) in the own setup scale with a sensitivity of ± 2 Pa and reproducibility of the measurements no worse than ± 5 Pa. The experimental setup in [98, 99] used almost the same parameters.

Such setups are extremely rare in the world and the quality of the data obtained is usually well known. The (pVT)-setup with visual observation discussed here is not an exception (see, e.g., [40], p. 112). In addition to subjective judgements on this theme, one can obtain an objective idea about the accuracy, sensitivity and reproducibility of the data in this experiment from Figs. 1.6, 1.12, and 1.21, where (partly or in full) the coexistence curve, critical isotherm, and critical isochore of pure SF_6, obtained with the help of this setup, are shown [87–91, 128]. Figure 1.6, for example, clearly shows that the scattering of experimental points lies within the margin of error for temperature. In Fig. 1.12, where only a part of the isotherm, immediately adjacent to the critical point is shown, apart from the horizontal part the precision of the pressure measurement and stability of thermostatting are distinctly visible. It should be noted that the "height" of the rectangle in the graph is twice the measurement error of pressure (which indirectly also includes temperature instability, which gives $\sim 1/3$ of the whole of error when $(dp/dT) \sim 8 \times 10^5$ Pa K^{-1} which does not exceed 0.5 g at 38 kg (calculated at 1 cm^2) that is better than $\sim 0.001\%$. As for the peculiarities of Wagner and colleagues' (pVT)-setup one can find more details in [98, 99].

1.3
Experiments Near the Critical Point in the Presence of the Gravitational Field

In this section we shall discuss the results of precise (pVT)-experiments, in which the nearest vicinity of the critical point of pure sulfur hexafluoride was studied in detail. The coexistence curve, critical isotherm, isothermal compressibility in the single-phase area, and the (pT)-dependence along the critical isochore were investigated [87–92, 126–128]. All the experimental material was concentrated in narrow temperature (-0.3 K $< T - T_c < 1.3$ K) and density ($|\Delta\rho^*| \leq 0.15$) intervals. The main result of this research, apart from the high accuracy experimental data themselves, was, as we have already said, in obtaining three static critical indices (β, γ, and δ) under conditions of the same experiment on the same experimental setup. It became clear that in the studied vicinity of the critical point there are own "far" and "near" regions. In the far region, all the critical indices have values close to Ising, whereas in the "near" one the indices again become typical for mean-field behavior. In the works [87–91] such a behavior was for the first time related to the strong specific influence of gravitation. Such an unusual, for its time, interpretation has still not become generally accepted up to now, which makes it necessary for us, even if briefly, to recall the main facts.

1.3.1
The Gravitational Effect

As for the experiment, the effect of the gravitational field, as has already been mentioned, on critical phenomena was first noted and described surprisingly accurately and comprehensively more than 100 years ago by Gouy [28]. He was the first to describe the density distribution of matter along the height of the vessel in the presence of gravitation under conditions of extremely high compressibility near the critical point. Noting the necessity of having a very high degree of temperature stability, he wrote: "changes by several thousands of a degree per hour completely impede observations which only become satisfactory if the changes do not exceed 0.0001°" ([28], quoted from [31], p. 334). Gouy was actually the first to connect the presence of a flat top of the coexistence curve near the critical point with the gravitational effect and noted the necessity of stirring in order to achieve equilibrium, which otherwise would not happen "even after one week had passed." Despite the brevity of Gouy's remarks it is possible to get quite a good idea of how well his experiment was carried out and the high degree of his own observational skills and intuition even though the experiment was performed in the 19th century.

A brilliant continuation of Gouy's work was carried out in the mid-1950s by a group of Canadian researchers [32–34] who studied the coexistence curve on a (pVT)-setup with visual observation and stirring. This analysis was the first quantitative study of the effect of gravity on the shape of this curve in the vicinity of the critical point. In this paper, they managed to achieve a record, for that time, measurement accuracy; temperature was determined with an accuracy of $\pm 0.001\,°C$, critical density with $\pm 0.2\%$, and pressure with $\pm 0.01\%$. Technically the research was also faultlessly designed. The gravitational effect on the shape of the curve in big (190 mm) and small (12 mm) vessels was studied. This was in fact a test of Mayer's theory [134], which it did not hold up and so ceased to exist. Today the Mayer theory, according to which the coexistence curve has a flat top even without the gravitation effect, has been almost forgotten.

Leaving aside the already familiar discussion about gravitation as the reason for the coexistence curve's flat top let us look at the clear, but up to now, nobody noted asymmetry of the gravity field effect on the different branches of this curve. The effect of the field on the liquid branch stops being noticeable, i.e., both curves close up, when $|\tau| = 1.38 \times 10^{-3}$, while the merging of the gas branches only happens when significantly removed from the critical point by $|\tau| = 4.8 \times 10^{-3}$ (see Fig. 1.3). While looking at the form of the critical isotherm [88, 90, 91] (see Chapter 2) we will see that the behavior of its "gas" and "liquid" branches is analogous to the case considered. We do not know the reason for such peculiarity.

When discussing the gravitational effect we must not forget to mention the influence stirring has on it. Many years after Gouy, D. Cannell from the University of California studied this problem and came to the fundamentally important conclusion [135] that stirring, which removes the density gradients caused by

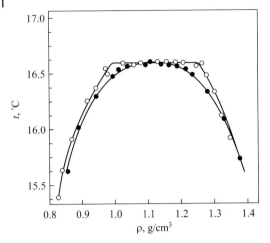

Figure 1.3 The coexistence curve of Xe in vessels of big (19 cm) and small (1.2 cm) height [32].

gravitation, does not affect the critical properties of the studied matter. It should be added that the experiments discussed [98, 99] were carried out without stirring and led to exactly the same results as in the experiments with stirring [84, 87, 88, 91, 92]. The only difference was that it took several hours to achieve equilibrium in the first case while it only took a few minutes when stirring took place. It should also be mentioned that stirring can help to get rid of the consequences of the "primitive" gravitational effect only, which consists in the redistribution of density along the vessel height, and apparently does not affect the manifestation of the intrinsic gravitational effect which modifies the nature of the phase transition (for more see the second paragraph in Chapter 2 and [181–183]). We believe that this is the reason why our results coincide with those of Wagner and colleagues independently of the presence or absence of stirring. Therefore, when discussing the gravity influence on a measurement of a property in the nearest vicinity of the critical point, we will mean only the intrinsic gravitational effect.

1.3.2
The Coexistence Curve

The equation for the determination of the coexistence curve, according to extended scaling (see, e.g., [63, 104]), is

$$\Delta\rho^* \equiv \frac{\rho_{\text{liq}} - \rho_{\text{vap}}}{2\rho_c} = B_0(-\tau)^\beta \left(1 + B_1 |\tau|^\Delta + \cdots\right). \tag{1.4}$$

Here, ρ_{liq} and ρ_{vap} are the orthobaric densities of the coexisting liquid and vapor, respectively; ρ_c is the critical density; B_0, B_1, and β are the critical amplitudes and critical exponent of the coexistence curve; and Δ is the so-called Wegner exponent, a universal correction of the main asymptotic term [63]. Calculation using the ε-expansion gives $\Delta \approx 0.5$ (more precisely 0.493, see, e.g., [103]). Other possible representations of a coexistence curve look like

$$\left.\begin{aligned}\Delta\rho_+^* &\equiv \frac{\rho_{liq} - \rho_c}{\rho_c} = B_{0+}(-\tau)^{\beta_+}\\\Delta\rho_-^* &\equiv \frac{\rho_c - \rho_{vap}}{\rho_c} = B_{0-}(-\tau)^{\beta_-}\end{aligned}\right\}, \tag{1.5}$$

where B_{0+} (β_+) and B_{0-} (β_-) are the critical amplitudes (critical exponents) of the liquid ($\rho > \rho_c$) and vapor ($\rho < \rho_c$) branches of the coexistence curve, respectively. On the coexistence curve, $\tau < 0$ holds.

Note that Eq. (1.4) is of more traditional nature. It implies that the critical behavior of real systems coincides with an idealized model behavior for which the equality $\beta = \beta_+ = \beta_-$ is characteristic. The two other representations given by Eq. (1.5) are useful, because they allow one to reveal the possible asymmetry of the upper part of the coexistence curve, which is manifested in the inequality between the critical exponents β_+ and β_-.

The coexistence curve is fairly simple to study as it is not necessary to measure the most "labor intensive" parameter, pressure. Density is measured by optical or dielectric methods using the Lorentz–Lorenz or Clausius–Mossotti formulas. A large number of papers have been dedicated to research on this curve (see, e.g., [39, 40, 49, 65, 66]). However, the results of these studies cannot be easily used for our purposes as they were not carried out very near to the critical point. In the case of optical measurements this restriction is caused by the need to avoid considering multiple scattering when working in the region of developed critical opalescence, where the laser beam deviates from rectilinear propagation and then completely disappears, totally scattering in the cell filled with matter (see Chapters 5 and 6 for more details).

Dielectric experiments have their own difficulties. For example, the authors of the carefully carried out investigation [136] had to ignore all experimental points, located in the region $|\tau| < 4 \times 10^{-4}$, in their analysis because, as they said, of the distorting effect of gravitation. At the same time, as the papers [87, 89, 91] showed, the peculiarities in the coexistence curve's behavior caused by gravitation begin only (!) when $|\tau| \leq 10^{-5}$. The natural desire to reduce the gravitational effect in dielectric experiments makes it necessary to use even thinner condensers where the effect of surface forces can be fully compared to the effect of gravitation (see, e.g., [137]). To our opinion, it is the surface forces, but not gravitational ones, in capacitors with a very small gap (0.076 mm [136]) that are the real reason for the observed effects in this paper in the region $|\tau| < 4 \times 10^{-4}$.

It should also be mentioned that both optical and dielectric methods are indirect and their applicability in such complex conditions is not very obvious. It is well known that to apply the Lorentz–Lorenz formula one must adequately account for local field effects, which is not very simple to do (see Chapter 7 for more). This problem was especially studied in [138, 139] in which Beysens et al. concluded that this formula could not be applied near the critical point. The Clausius–Mossotti approximation suffers from the same inadequacies as the Lorentz–Lorenz formula as they are practically two different formulations of the same idea. Thus, by attempting to apply these dependences in order to study critical phenomena, the researcher finds himself in a kind of closed circle. This is because the main peculiarity of these formulas is that they are very approximate everywhere [140], especially in the vicinity of the critical point. If the critical point is excluded from the study then, as experience shows, both methods lead to very reasonable results [40].

In contrast, precise (pVT)-experiments with visual observation and stirring are free of all the listed shortcomings and, to our opinion, are perfectly suitable for studying static properties near the critical point. However, there is a very widespread opposite view that the (pVT)-method is limited by the gravitational effect (see, e.g., [40]) which disturbs the experiment in the immediate vicinity of the critical point, deforms the shape of the curve, distorts the critical indices etc., and stops them being "true."

Actually, if one accepts an alternative point of view, stating that the critical indices are not distorted by gravitation but obtained in its presence and, consequently, are true for these circumstances, then the situation is cardinally changed. This approach also gives us a way of looking deeper into the nature of critical phenomena [2, 141, 142]. All known experimental material shows that the presence of an additional factor such as the weak influence of the gravitational field (in small vessels) really makes it possible to reveal new features of critical behavior (concerning the effect of the weak magnetic field on phase transitions near the Curie point; see, for example, Chapter 4 in [143]).

In (pVT)-experiments [87,91] more than 60 experimental points on the coexistence curve of pure SF_6 in the range $\Delta T = 0.2$ K below T_c were registered (Figs. 1.4 and 1.5). Moreover, the top of the curve was studied in detail. The points, closest to the critical state, are only 0.0001 K apart from it, while, in the interval $\Delta T = 0.003$ K, 10 values were obtained on the "liquid" branch and 11 on the "gas" branch (Fig. 1.6). In fact, this interval is 15 times larger than the error in temperature measurements.

One can see the considerable asymmetry of the curve's top. It is also found that the critical density value ($\rho_c = 0.73883$ g cm^{-3}), obtained by visual observation, significantly differs from the value $\rho'_c = 0.7416$ g cm^{-3}, arrived at using the "rectilinear diameter" rule. All the data and figures presented hereinafter reflect the results of a reanalysis [141] of experimental data [87, 89, 91]. Modern computer analysis techniques made it possible to bring in correspondence calculation accuracy with volume measurements precision. Previously, the determined ρ'_c-value was $\rho'_c = 0.742$ g cm^{-3} [91]. The mentioned difference, as already discussed, is 0.4%, which is 20 times higher than the error of volume measurements. In our opinion, there is nothing very surprising about this difference as the critical density determination was practically reduced to the measurement of the coexistence curve part closest to the critical point [87, 91], as was suggested in [84]. Thus, the asymmetry of the whole upper part of the coexistence curve finds its the natural reflection in the critical density shift. The reasons for this asymmetry, as was already noted, can be attributed to gravitation.

The coexistence curve is shown in Fig. 1.7 in a wider temperature range $\Delta T \geq 0.03$ K, together with its diameter which, starting from $\Delta T = 0.01$ K ($\tau \sim 3 \times 10^{-5}$),

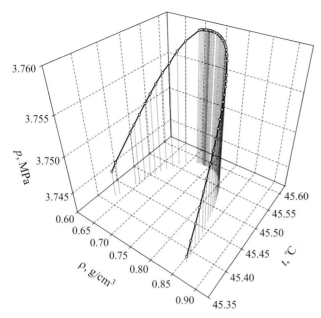

Figure 1.4 Three-dimensional representation of the coexistence curve of SF_6. The domain of variability of state parameters is $(T_c - T) \approx 0.2\,K$, $p_c - p = 17\,kPa$, $\Delta\rho^* = \pm0.15$ [141]. The plots are drawn using data from [87, 91].

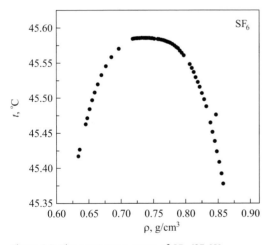

Figure 1.5 The coexistence curve of SF_6 [87, 91].

becomes curvilinear [87, 89, 91]. Such behavior in pure liquids was simultaneously and independently found on SF_6 [87, 144]. In [87], it was observed using direct (pVT)-measurements, and in [144] by analyzing the temperature dependence of the dielectric constant. These results not only aroused a wide discussion at the time, but also led to a great number of additional experiments dedicated to this

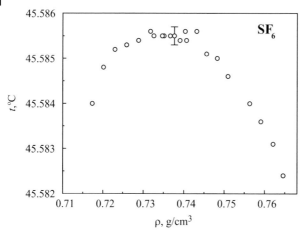

Figure 1.6 The top of the coexistence curve of SF_6 in the temperature range $(T_c - T) = 0.003$ K. The size of the vertical line corresponds to the temperature measurement error $\pm 2 \times 10^{-4}$ K [87, 91].

problem (see, e.g., [98, 121, 125, 129, 131, 145–151]). It should be said that during all these researches the existence of a singular diameter in nonconducting liquids was not experimentally confirmed, and moreover, no single reason for its possible appearance was proposed.

In Fig. 1.8, the data for the coexistence curve for pure SF_6 (see Figs. 1.4 and 1.5) are shown in the log–log scale. As all values of $\Delta\rho^*$ (see Eq. (1.5)) for this graph were calculated using as critical density the values ρ_c', determined by the rectilinear diameter rule, the fact that the experimental points corresponding to both branches

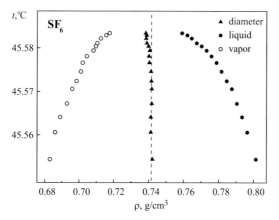

Figure 1.7 The top of the coexistence curve of SF_6 in the temperature range $(T_c - T) = 0.03$ K together with its "curvilinear diameter". For $(T_c - T) > 0.01$ K, $\rho_d = \rho_c' = 0.7416$ [87, 91].

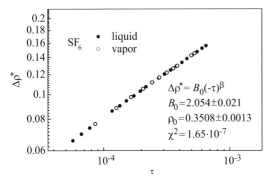

Figure 1.8 The coexistence curve of SF_6. The symmetry of the behavior of both branches of the curve in the interval $5 \times 10^{-5} < |\tau| < 7 \times 10^{-4}$ [141]. The curves are plotted using data from [87, 91].

of the coexistence curve form in the region $|\tau| > 3 \times 10^{-5}$ a common line definitely indicates that this rule is well fulfilled in this temperature range.

Figure 1.9 demonstrates the accuracy of the approximation of both branches of the coexistence curve in the interval $3 \times 10^{-5} < |\tau| < 7 \times 10^{-4}$ by simple scaling dependences (Eq. (1.5)) with $B_{0+} = B_{0-} = 2.054 \pm 0.021$ and $\beta_{+} = \beta_{-} = 0.3508 \pm 0.0013$ [141]. It should be mentioned that these refined values agree wonderfully with the results of the "pre-computer era" coexistence curve analysis $B_0 = 2.05 \pm 0.01$ and $\beta = 0.350 \pm 0.006$ [89, 91, 126, 128]. Note again that Eq. (1.5) describes each branch of the coexistence curve separately in contrast to the more usual equation (1.4). From Fig. 1.9 it follows that in the "far" part of the mentioned temperature range errors are random and do not exceed 0.02–0.03%, while in the "near" part there can be observed small deviations ($\sim$0.06–0.07%) but systematic ones. As analysis showed, the addition of extra terms (such as a dependence like Eq. (1.4)) does not lead to a better description of the experimental data. A very detailed analysis of this problem was especially carried out in [153]; it was dedicated to the analysis of xenon. It was shown that the obtained coexistence curve data could be well described by the simplest one-term equation with a common index $\beta = 0.356$ for the whole temperature interval $2 \times 10^{-5} < |\tau| < 3 \times 10^{-2}$ studied. Moreover, applying in this case Eq. (1.4) in its full form could not, as well as for SF_6, improve this description. This result clearly shows that it is not necessary to apply the extended scaling for such small temperature intervals (see also [125]).

At the same time it is clear (Fig. 1.9) that the character of deviations near the critical point ($|\tau| > 3 \times 10^{-5}$) is such that they could probably be avoided by using the same small correction, adding it to the values on the liquid branch of the coexistence curve and subtracting it from the gas branch values. Opposite correction signs, here and in the area of strong asymmetry of the coexistence curve ($|\tau| < 10^{-5}$), can, in our opinion, be explained by the influence (naturally, weaker) of the gravitational effect. It is clear from the diagram that for SF_6 (vessel height 8 mm) the gravitational effect on the coexistence curve disappears completely,

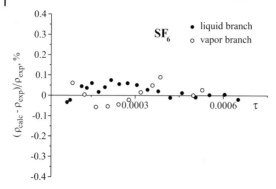

Figure 1.9 Coexistence curve of SF_6: reduced deviations of calculated density values (Eq. (1.5)) from experimental ones [141]. The curves are plotted using data from [87, 91].

beginning from $|\tau| > 4 \times 10^{-5}$. Considering the results for the different heights of the vessels used in two other papers [32, 98], where the gravitational effect was also found, this value agrees well with the "old" data [32] as well as with the "new" results [98].

1.3.3
Singularity of the Diameter of the Coexistence Curve

More than 100 years ago, Cailletet and Mathias experimentally established the so-called rectilinear diameter rule for the coexistence curve

$$\rho_d \equiv \frac{\rho_{\text{liq}} + \rho_{\text{vap}}}{2\rho_c} = 1 + A_1|\tau| + \cdots. \tag{1.6}$$

This rule remained correct even when the precision of the experiment and the degree of approach to the critical point already made it possible to find nonclassical values of the critical index β (see, for example, reviews of experimental work performed in the mid-1970s [39, 49], where the singular diameter problem did not even arise, or the papers of the same time, where this problem was especially looked at and it was shown that the experimental data for all substances, investigated by that time, agreed well within the margins of error with the rectilinear diameter rule). According to our data, this rule is well satisfied right up to $|\tau| \geq 3 \times 10^{-5}$ (see Figs. 1.7 and 1.8). In [98], which devoted particular attention to this problem, deviations from the rectilinear diameter rule were not found and it was used to determine the critical density.

There also appeared theoretical papers at the beginning of the 1970s, which suggested the possibility of incomplete symmetry between liquid and its vapor (particle–hole systems in terms of the lattice model) and consequently this rule was violated (see, e.g., [119, 120, 155, 156]). The singularity is not actually observed on the diameter itself but on its first derivative. It should be mentioned that these papers were only dealing with theoretical models. As a result, "singular" term was added to Eq. (1.6):

$$\rho_d^* \equiv \frac{\rho_{\text{liq}} + \rho_{\text{vap}}}{2\rho_c} = 1 + A_0|\tau|^{1-\alpha} + A_1|\tau| + \cdots, \tag{1.7}$$

where α is the critical index of isochoric heat capacity.

Application of the RG-representation to explain this problem in those cases, when the critical behavior of liquids was initially modeled with the addition of an asymmetrical term to the ϕ^4-Hamiltonian [149,150], really led to the appearance of a nonanalyticity like Eq. (1.7). In other cases, a rectilinear diameter was obtained [157]. In [121–124], which further developed this theme, the singular diameter arose by taking into account not only double interactions but also triple ones. It seems that for the first time this approach was applied in [121] as an immediate reaction to the results of our research [87]. Without going into details, it is worth mentioning one important circumstance: the gravitational effect was not taken into consideration in any of these theoretical investigations.

As for the paper [144] where, according to its authors, the singular diameter was first found on the coexistence curve of SF$_6$, we shall add just one remark. Despite the wide resonance this paper arose in that time (see, in particular [148–150, 158], the deviation of the diameter from rectilinearity (see Fig. 1.10) was found in this work so far from the critical point ($\Delta T = 0.3\,^\circ$C), where nobody before or since has observed such effects (see, e.g., [87, 98, 99, 151] and Figs. 1.6–1.8). In this connection, the detailed investigation [151] deserves special attention, since here the same substance (SF$_6$) was studied using the same dielectric method as in [144], but with greater precision (it is enough to mention that the thermostatting stability in this paper was $\sim$20 μK (!), against $\sim$0.001 K in [144]). No deviations from rectilinearity were found and, according to the authors [151], the gravitational effect began to have a strong influence on the coexistence curve only in the range $|\tau| < 1 \times 10^{-4}$. It should be mentioned that with a height of the measuring cell (capacitor) of 0.5 mm it is impossible here as well as in [136] to completely exclude the influence of the surface tension forces. However, neither in [144] nor in any later work, including the masterly analysis carried out in [148] in order to prove that in [144] it was the singular diameter that was observed, the gravitation as a possible reason for the observed deviations was not even considered. Moreover,

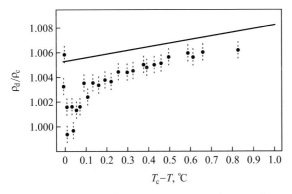

Figure 1.10 Behavior of the coexistence curve diameter for SF$_6$ [144].

the results of our paper [87], despite the suggested explanation related just to gravitation, were considered by other authors as direct experimental evidence of the diameter singularity of the coexistence curve (see, e.g., [121, 125]). Now, however, after the appearance of the papers [98, 99], there can be no doubt in our mind that gravitation is a cause of curve coexistence asymmetry. Without denying the existence of the diameter singularity for corresponding theoretical models [119, 120, 155, 156], we shall present below (see Chapter 2) a possible alternative explanation for the coexistence curve asymmetry, not for a model system but for a real liquid, where the gravitational field will be given the main role [87, 89, 91] (see Fig. 2.4).

1.3.4
The Critical Isotherm

Out of all the lines, which are usually investigated when looking at critical phenomena, the critical isotherm is of particular importance due to its most complex and labor-intensive experimental study. An indirect confirmation of this could be the fact that in vast reviews [39, 40] it was not practically discussed, while the coexistence curve had a separate paragraph dedicated to it and susceptibility (compressibility) a large table. This is a reflection of real difficulties, which arise when investigating the critical isotherm, whose direct (pVT)-measurements are extremely rare [88, 98, 159].

The most detailed critical isotherm of SF_6 [88] ($\sim$140 experimental points in the range of density change, $\Delta\rho^* = \pm 0.2$) is shown in Fig. 1.11. In the two other cited works on the critical isotherm of SF_6 [98] and Xe [159], about 25–30 experimental points were obtained. Figure 1.12 shows only a part of this isotherm immediately

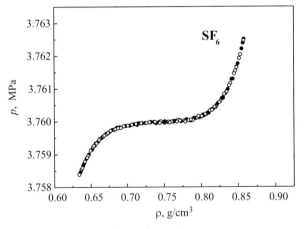

Figure 1.11 Critical isotherm of SF_6. The measurements were carried out at increasing (•) and decreasing (○) volume [88]. The absence of hysteresis shows that experimental data are obtained for equilibrium.

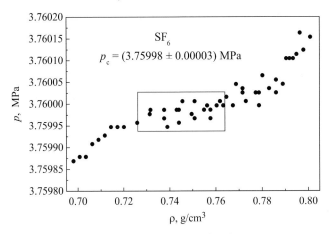

Figure 1.12 The horizontal part of the critical isotherm around the critical point: new analysis [141] of experimental data given in [88].

close to the critical point. Here, apart from the horizontal part, both the precision of the pressure measurement and thermostatting stability are clearly seen. The presence of a horizontal part of the isotherm and the flattening of the upper part of the coexistence curve are consequences of the gravitational effect [32, 115]. From the data it can be seen that its width on the critical isotherm ($\sim\pm2\%$) is wider than that on the coexistence curve ($\sim\pm1\%$). In our opinion this can be easily explained, as the "moving away" from the critical point along the coexistence curve takes place simultaneously with two parameters (ρ and T), and for the isotherm only with one parameter (ρ). It is therefore natural that gravitation in the second case is effective in a wider range of density changes than in the first case. The fact that there is a twofold difference in the dimensions of the flat parts on these curves does not seem to be an accidental coincidence and may also point to the gravitational nature of the horizontal part of the isotherm.

The experimental data relating to the critical isotherm [88] were treated in [90, 91, 126, 141] in terms of the similarity hypothesis both in ($\mu - \rho$)- and ($p - \rho$)-representations in accordance with Eqs. (1.8) and (1.10):

$$\Delta\mu_{\pm}^{*} = D_0 \Delta\rho_{\pm}^{*} \left|\Delta\rho_{\pm}^{*}\right|^{\delta-1} \left(1 + D_1^{\pm} \left|\Delta\rho_{\pm}^{*}\right|^{\Delta_{\pm}/\beta}\right), \tag{1.8}$$

where μ is the chemical potential,

$$\Delta\mu^{*} \equiv \frac{\rho_c}{p_c} \left[\mu(\rho, T) - \mu(\rho_c, T_c)\right], \tag{1.9}$$

and the signs "+" and "−" refer to the liquid and vapor branches of the critical isotherm, respectively,

$$\Delta p_{\pm}^{*} = D_0 \Delta\rho_{\pm}^{*} \left|\Delta\rho_{\pm}^{*}\right|^{\delta-1} \left(1 + D_1^{\pm} \left|\Delta\rho_{\pm}^{*}\right|^{\Delta_{\pm}/\beta}\right),$$

$$\Delta p_{\pm}^{*} \equiv \left(\frac{p_{\pm}}{p_c} - 1\right). \tag{1.10}$$

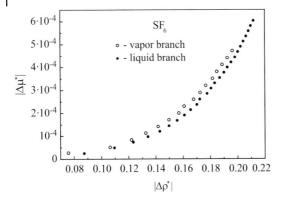

Figure 1.13 Comparative position of both branches of the critical isotherm of SF$_6$ [88, 90, 91].

From a theoretical point of view, the $(\mu - \rho)$-representation is more adequate as it is suggested that in this case the critical isotherm should be fully antisymmetrical. In experiments, the critical isotherm is measured in more natural $(p - \rho)$-coordinates. Wallace and Meyer, the authors of an extended research of static properties of ^{3}He near the critical point by the dielectric method [161, 162], expressed their surprise after they carried out numerical integration of experimental pressure values on the critical isotherm and were unable to obtain an equality of chemical potential values on its both branches, as was suggested by contemporary theory (see, e.g., [39]). Moreover, it turned out that $|\Delta\mu^*_{gas}| > |\Delta\mu^*_{liq}|$. To find this experimental fact requires high sensitivity and precise data, which does not actually happen very often. In the experiment on SF$_6$ (Fig. 1.13), the same result was obtained [88, 90, 91].

Concerning the discovered deviation from complete antisymmetry of the critical isotherm, the calculated $(\mu - \rho)$-data do not, to our opinion, have enough advantage over the initial $(p - \rho)$-representation. Therefore, the new analysis [141] of the critical isotherm of SF$_6$ was only carried out with Eq. (1.10). It turned out that just as in the case of the coexistence curve, to obtain the same critical index (δ) and amplitude (D_0) for both branches of the critical isotherm it was necessary to use, instead of the critical density value found by visual observation, the value which was determined by the "rectilinear diameter" rule. This gives an additional evidence of its validity.

The parameters of the main term (D_0 and δ) coincided with those for the $(\mu - \rho)$-representation found previously [90, 91], which are $D_0 = 1.70 \pm 0.01$ and $\delta = 4.30 \pm 0.01$. This result should have been, in principle, expected as the difference in the numerical values of $\Delta\mu^*$ and Δp^*, at worst, does not exceed $0.01\Delta p^*$. Moreover, due to the carried out analysis it was possible not only to fulfill the requirement of the equality $\Delta_+ = \Delta_- = \Delta$ [63], which had been already obtained earlier [90, 91], but also to achieve a quantitative coincidence of the obtained values $\Delta = 0.49 \pm 0.02$ with its RG-value, $\Delta = 0.493$ [103]. Thus, the full set of parameters of Eq. (1.10) for describing the critical isotherm of SF$_6$ "far" from the critical point

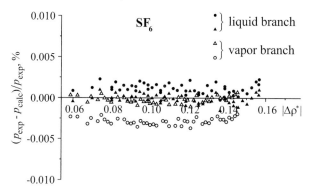

Figure 1.14 Critical isotherm of SF$_6$. The relative deviations of the calculated pressure values (Eq. (1.10)) from experimental ones [88]: $p_c = 3.75998$ MPa for both branches of the critical isotherm ($\bullet$, $\circ$); $p_c = 3.75988$ MPa for the gas ($\triangle$) branch and $p_c = 3.76003$ MPa for the liquid ($\blacktriangle$) branch [141].

($|\Delta\rho|^* > 0.08$) turned out to be as follows:

$$D_0 = 1.70 \pm 0.01, \qquad \delta = 4.30 \pm 0.01, \tag{1.11}$$

$$\Delta_+ = \Delta_- = \Delta = 0.49 \pm 0.02,$$

$$D_1^+ = 1.40 \pm 0.05, \qquad D_1^- = 0.06 \pm 0.02,$$

$$\beta = 0.3508 \pm 0.0013.$$

The comparison of experimental and calculated, according to Eq. (1.10), taking into account Eq. (1.11), pressure values is presented in Fig. 1.14. Looking at this picture, we can see systematic deviations, although small and within the limits of $\pm(0.002 - 0.004)\%$, especially on the gas branch. We attribute this critical isotherm asymmetry, as well as the coexistence curve asymmetry (see Figs. 1.6 and 1.7), to gravitation. The asymmetry observed in both cases cannot be explained by a real difference between liquid and gas, as any difference between them should be decreased on approaching the critical point. As the analysis in [141] showed, the observed systematic deviations can only be removed by a special choice of the critical pressure. No changes in the critical indices and amplitudes led to the desired result (Fig. 1.14). Just like in the case of the coexistence curve (see Fig. 1.9) it is necessary to add to the measured pressure value (there, density) on the liquid branch and subtract from the gas branch some small correction which gets smaller as the system moves away from the critical point. It is remarkable that if the error in measurement of pressure, including critical, was higher by only two or three times, it was impossible to see the described peculiarities of the critical isotherm. Even in this case the measurement precision would have been better than 0.01% which is much higher than in most of the researches that we are aware of.

Finally, just as with the analysis of the coexistence curve, in the nearest vicinity of the critical point the change in the critical index toward its classical value was observed. The change here was also attributed to gravitation. Analogous results with the same interpretation were also presented in [98] (for a detailed and comparative analysis, see Chapter 2).

1.3.5
Isothermal Compressibility Along the Critical Isochore

This thermodynamic characteristic being a particular case of a more general property, the system susceptibility, plays a rather noticeable role in the physics of critical phenomena as well as in the physics of condensed state as a whole. It has been known since van der Waals that all the phenomena, united by the term "critical," are directly related to singularity of the compressibility. This fact is also reflected in the variety of methods for its experimental determination using the temperature dependences of different quantities like the intensity of scattered light and X-rays, density, dielectric constant, etc. (see, e.g., [39, 40, 49, 65, 66, 125]). As the first and, for a long time, as the only one Habgood and Schneider's work [159] devoted to the thorough and detailed (pVT)-analysis of the critical region of xenon should be considered. The authors did not use their experimental data to find the compressibility critical index γ. This was, however, done subsequently by Kadanoff et al. [48] using the original data from [159] for the determination of the dependence of compressibility on temperature in accordance with Eq. (1.12):

$$p_c K_T \equiv \frac{p_c}{\rho_c} \left(\frac{\partial \rho}{\partial p} \right)_T = \Gamma_0^+ \tau^{-\gamma}. \tag{1.12}$$

The result is shown in Fig. 1.15: "far" from the critical point the value of γ was equal to 1.44, which is close to its Ising value, near the critical point $\gamma \sim 1$ holds, which is typical for the mean-field behavior. A similar result was obtained 20 years later on SF_6 [92, 163]. The paper [92] was a direct continuation of the research carried out in [163] and its intention was to extend this research by studying more carefully the nearest vicinity of the critical point. Although the experiment's precision, as described in [163], gave us no reason to doubt the validity of the obtained values of the critical index of compressibility, due to the unusual importance of this problem for critical phenomena further investigation of the temperature dependence of compressibility was carried out in the temperature interval $(0.02 < (T - T_c) < 0.33 \text{ K})$, where in [163] was found $\gamma = 1.00 \pm 0.02$ (Fig. 1.17).

With this aim the same experimental setup as in [163] was used to study 16 isotherms (Fig. 1.16) lying higher than the critical one with an extremely small temperature step (~ 0.02 K). On each isotherm ~ 20 experimental points were obtained within the range of volume changes of $\pm 8\%$ from the critical one with a decrease in the substance's density in the piezometer as well as with its increase. The absence of hysteresis (see Fig. 1.16) shows that the obtained data

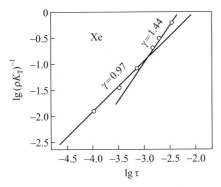

Figure 1.15 Data on the isothermal compressibility of xenon [159] as processed by Kadanoff et al. [48].

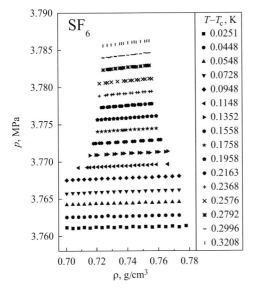

Figure 1.16 Linear sections of the $(p - \rho)$-isotherms in the single phase region of pure SF_6 in the temperature interval $0.02 < (T - T_c) < 0.33$ K [91, 92].

were measured for equilibrium states of the system under consideration. The data analysis carried out in [89, 91] confirmed the result obtained in [163] and showed that there are two regions in the vicinity of the critical point: "far" $(\tau > 10^{-3})$, where the critical index $\gamma = 1.16 \pm 0.03$ turned out to be close to the Ising value $\gamma = 1.24$, and "near" $(\tau < 10^{-3})$, where $\gamma = 0.96 \pm 0.02$ (see Fig. 1.17), which characterizes mean-field behavior. This does not exclude the fact that it should be looked at as a crossover from Ising to mean-field value of the critical index of compressibility.

The results obtained in [89, 91] and in [163] were fully identical, but the principal difference between these works is connected with their interpretation. The authors

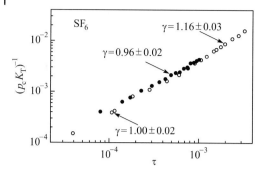

Figure 1.17 Dimensionless inverse isothermal compressibility dependent on the approach to the critical point (• [89, 91], ○ [163]).

of [163] suggested that such behavior was a natural consequence of the classical theory of critical behavior. Another point of view [89, 91] suggests that it is the gravitational effect which leads to a change in the index γ to its classical value in the discussed experiments. In fact, we deal here with that gravitational effect that could not be eliminated by stirring (see, e.g., [135]). It was subsequently called the *intrinsic gravitational effect* [181–183].

Subsequently, in the 1990s Wagner et al. obtained exactly the same result ($\gamma = 0.98 \pm 0.05$ at $\tau < 5 \times 10^{-4}$) and again on SF_6 [98] (it is surprising but historically true that all nontrivial results on the behavior of simple liquids near the critical point were obtained for the first time on SF_6; of course, this statement is true except of the first result, which discovered the critical point itself, where CO_2 was used [7, 8]). Later on, apparently in order to convince themselves of the universality of such behavior, Wagner's group carried out detailed comparative research on the isothermal compressibility behavior of SF_6 and CO_2 in vessels of different heights [99]. In fact, it was a set of the same vessels placed, as in the famous research [32], either horizontally ($h = 11$ mm) or vertically ($h = 30$ mm). This work showed that the change, in the presence of gravitation, of the compressibility critical index from the one close to the Ising value $\gamma \approx 1.2$ to the mean-field one, $\gamma \approx 1$, took place both for SF_6 and CO_2 (see Figs. 1.18 and 1.19) and, in all probability, represented a universal property of pure nonconducting fluids. Moreover, in this work as well as in [89, 91, 92] it was possible to confidently determine the temperature of transition from one γ value to another. The temperature values obtained in [99] ($\tau \sim 4 \times 10^{-3}$, $h = 11$ mm) agree well with those found in [91, 92] ($\tau \sim 1.2 \times 10^{-3}$, $h = 8$ mm). It also turned out that the difference, found in [99], in the distance of this transition from the temperature critical point for both heights and both substances (see Figs. 1.18 and 1.19) completely agree with their density difference:

$$\frac{\tau_{CO_2}}{\tau_{SF_6}} \simeq \frac{\rho_{CO_2}}{\rho_{SF_6}} \simeq \frac{2}{3}. \tag{1.13}$$

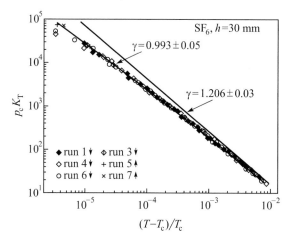

Figure 1.18 Compressibility of SF$_6$ [99].

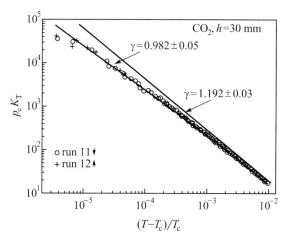

Figure 1.19 Compressibility of CO$_2$ [99].

It is difficult to disagree with the authors of [99] who suggested that this result is a very weighty argument in favor of the decisive role that gravitation plays in this effect.

Overall, these facts once more clearly show that, under conditions of unlimited growth of the system's compressibility as it approaches the critical point, sooner or later necessarily there will come such a moment when small, in the usual sense, gravitational effects become essential that lead to equally substantial consequences, namely, to the transformation of the fluctuation type of critical behavior to the mean-field, classical one. This effect was observed in [87, 89, 91, 92, 98, 99, 163] when not only compressibility but also all the other investigated thermodynamic characteristics of matter near the critical point were analyzed (see Chapter 2 for more on this).

1.3.6
(p − T)-Dependence Along the Critical Isochore

For us the $(p − T)$-dependence is noteworthy mainly because its form makes it possible to judge the isochoric heat capacity (c_v)-behavior near the critical point and, therefore, the value of the critical index, α. This possibility is based on Eq. (1.14) which can be easily obtained by double temperature differentiation of Gibbs's thermodynamic potential

$$\frac{\rho c_v}{T} = \left(\frac{\partial^2 p}{\partial T^2}\right)_\rho - \rho \left(\frac{\partial^2 \mu}{\partial T^2}\right)_\rho. \tag{1.14}$$

It is expected (see, e.g., [26, 40]) that the second temperature derivative of the chemical potential has not a singularity at the critical point, so only $(\partial^2 p/\partial T^2)_{\rho_c}$ is responsible for the heat capacity singularity. Several experimental attempts were undertaken to use this equation for determining the value of the critical index α by experimental $(p − T)$-dependence along the critical isochore [91, 132, 152]. In Kierstead's work [132], the $(p − T)$-dependence along the critical isochore of ^{4}He was analyzed with very high accuracy. Meanwhile, to avoid using double differentiation, the derivative $(\partial p/\partial T)_{\rho_c}$ was directly measured. The ^{4}He studied in this work was high purity grade (99.999%), and its critical parameters were defined independently on the same experimental (pVT)-setup.

The authors of [152] in their extensive $(−6\,\text{K}< (T − T_c) < 46\,\text{K})$ research on CO_2 also used the (pVT)-method with an experimental setup like in [159], but without the possibility of visual observations. The accuracy of these measurements was considerably lower than in [132]. Data analysis, carried out in these works, actually led to the same result: $−0.2 \le \alpha \le 0.2$ [132] and $0 \le \alpha \le 0.15$ [152]. In other words, it turned out that neither of these attempts was, unfortunately, successful.

Our (pVT)-investigation was carried out on pure (99.9995%) SF_6 in [91, 92]. The motivation for research was the following. First, the experimental setup accuracy was significantly higher than that in [152]. Second, as for the negative result obtained for ^{4}He, then it, a quantum liquid, could, in principle, have a special behavior. What is more, if success is achieved, then in the same experiment the full set of critical indices $(\alpha, \beta, \gamma, \text{and } \delta)$, which are needed to describe static critical phenomena, could be simultaneously obtained. In the course of this experiment [91], within a fairly narrow temperature interval near the critical point $(−7 \times 10^{-4} \le \tau \le 3 \times 10^{-3}) \sim 220$ experimental points were found (Fig. 1.20). Particular attention was paid to studying the region $\pm 0.005\,\text{K}$ from the critical point. More than 40 pressure values for different temperatures within this range were registered (Fig. 1.21). Ideal linearity of the dependence near zero makes it possible to state that here, just as in [132, 152], $\alpha = 0$.

It would be natural to suggest that, just as for other critical indices, the obtained value of α is a consequence of the gravitational effect. It is known that gravitation distorts the heat capacity anomaly (see, e.g., [40]), which makes it similar to a jump that is typical for classical theory for which $\alpha = 0$. Thus, in the precise (pVT)-experiments described above [87, 89, 91, 92, 98, 99, 163] it was found every

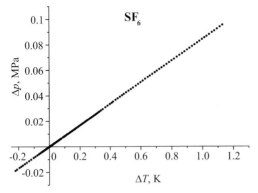

Figure 1.20 $(p-T)$-dependence along the critical isochore of SF$_6$ [91, 126].

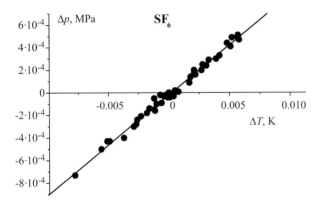

Figure 1.21 The nearest to the critical point part of the $(p-T)$-dependence along the critical isochore of SF$_6$ [91, 126].

time that critical indices took on their "classical" values, beginning with a definite degree of approach to the critical point with respect to temperature (density).

Let us repeat: the compressibility critical index turned out to be equal to $\gamma \approx 1$ for $\tau \leq 10^{-3}$, the coexistence curve critical index $\beta \approx 0.5$ for $|\tau| \leq 10^{-5}$, while on the critical isotherm $\delta \approx 3$ for $|\Delta\rho|^* < 0.08$. Such a type of behavior was first found in [87, 88, 91, 92], and for the first time explained by the strong specific influence of the Earth's gravitational field on different properties of matter under conditions of unlimited growth in the system's susceptibility. In [89] a nontrivial analogy in the critical behavior of pure liquids and magnets was noted consisting in the tendency of the index β to tend to its classical values common for both systems (see Chapter 2 for more). Also, the idea was put forward, which has only been confirmed in recent years, according to which "near to the phase transition points, independent of interaction details, external factors of different nature can in the same way influence the behavior of similar physical quantities" [89].

The mentioned paper can be considered as the first, maybe heuristic, formulation of this book's central idea: the nearest vicinity of the critical point is, in fact, the region of deformed (suppressed) fluctuations, whose further increase under conditions of unlimited growth in the system's susceptibility is restrained by the action of macroscopic fields of different (depending on the type of system) physical nature. As a result *the nearest vicinity of the critical point is transformed from a fluctuation region, where Ising model results are justified, to a region where classical, mean-field behavior should be once more re-established* [2, 126, 141, 142].

When this idea was first mentioned, just as a logical noncontradictory (at least, according to the author) suggestion [87, 89, 91], it had no other basis. Today the situation has changed significantly. And although everything that has been said above shows that, with rare exceptions, the problem of critical behavior of real systems in the presence of different disturbing factors has still not been definitely solved, recent experimental and theoretical researches, some of which we have already mentioned, add quite weighty arguments in favor of this idea.

2

Critical Indices and Amplitudes

2.1
Phenomenological Model of the Critical Behavior of Nonideal Systems

In the present chapter, a simple physical model based in fact on "first principles" of the theory of critical phenomena is proposed to explain the experimentally discovered nontrivial behavior of real systems near the critical point (see Chapter 1). Let the real physical system, which is already close enough to the critical point, get even closer to it, for example, by further changing the temperature (Fig. 2.1). Then, the order-parameter fluctuations in the system grow and, at some point, say at temperature T_1, become so developed that their radius (correlation length ξ) turns out to be, according to Kadanoff, the only determining scale characterizing the system's properties [61,62]. At this moment, the system goes over into region II and changes from mean-field behavior (region I) to fluctuation-determined behavior, in particular, to a behavior of Ising type. This transition can be called the *first crossover*. Its position is determined by the Ginzburg criterion [70]. For idealized systems it is assumed that such a change in the character of their behavior in the vicinity of the critical point can only happen once. The question now is whether the behavior of a real, nonidealized, system may change (and if it does, how) as it moves deeper into the fluctuation region.

It is well known that the development of large-scale fluctuations on approaching the critical point is accompanied by a continuous increase in the susceptibility (compressibility) of the "critical" system (see, e.g., [166]). This includes susceptibility to perturbations of different physical nature, such as gravitational and electric (internal or external) fields, surface forces and shear, samples nonideality and the presence of boundaries, turbulence and gradients, etc. Moreover, any disturbance, no matter how weak in ordinary circumstances it is, eventually becomes of outstanding importance here. This, in turn, means that any system in the nearest vicinity of the critical point should be considered as a nonideal system.

It should be remembered that by "nearest vicinity" we mean the vicinity of the critical point where the effect of one or another field in high precision and sensitivity experiments becomes noticeable. The following feature inherent to critical phenomena should be added to Kadanoff–Wilson's ideas that the critical point is a point of decreased stability [78]; in other words, the fluctuations become

Critical Behavior of Nonideal Systems. Dmitry Yu. Ivanov
Copyright © 2008 WILEY-VCH Verlag GmbH & Co. KGaA, Weinheim
ISBN: 978-3-527-40658-6

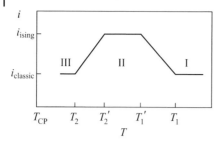

Figure 2.1 Curve, modeling the kind of behavior of critical indices (i) in real systems in dependence on their proximity to the critical point (CP) with respect to temperature (or density). For $i_{ising} < i_{classic}$ (for example, for the critical index β) the curve is mirror-symmetric with respect to the horizontal axis, passing through $i = i_{classic}$, and its flat part corresponds to $i = i_{ising}$ for this case.

extremely "fragile" formations, very sensitive to various perturbations. It therefore seems natural to assume that if we move deeper into the critical region, then the moment inevitably arises when fluctuations are initially deformed (at a temperature T_2', for example) and then (at a temperature T_2) are totally suppressed by one or other field applied to the system [2, 141, 142]. Even if other fields do not exist, as an ever-present factor the boundaries of the sample can play the role of such a field.

As a result, the system undergoes an inverse transition from fluctuation-type behavior (in particular, Ising type) to classical mean-field behavior and falls into region III (Fig. 2.1). By analogy, this transition can be called the *second crossover* [2, 141]. To judge whether one of these transitions (I⇔II or II⇔III) takes place, it is possible to use the well-known, but strictly individual for each of these regions, combination of critical indices. In such cases when the fluctuations of the order parameter in the system are still only developing (section $T_1 T_1'$ of region II), or they are already anisotropic but not completely suppressed by a field (section $T_2' T_2$), there should be a transitional type of behavior with intermediate critical indices. It should also be stressed that, in our opinion, the universality of critical phenomena consists, in particular, on the fact that in the limiting situations (see Fig. 2.1) no other critical indices apart from the fluctuational or classical ones probably seem to exist.[1] Indeed, the modern general theory of critical phenomena in idealized systems [71] gives, when the spatial dimensionality $d > 4$ (no matter what it is), mean-field, classical, indices while for any $d < 4$ it predicts Ising-type values [166]. The modern specific theory of the critical behavior of liquid in flows, in the presence of shear, based on the RG-approach [78] also gives in the limit the same two sets of index values (see below for more). It is interesting to note that the experiment [93, 95], in which it was quite easy to vary this type of perturbation, the entire spectrum of intermediate values of the critical susceptibility index (γ) between its known limiting values (see below and Fig. 2.14 for more) was found.

---

1) It is worth remembering in this respect Einstein's words: "Subtle is the Lord but malicious He is not."

By using Kadanoff's idea it is quite simple to understand the physical reason for this: if the system has the developed isotropic fluctuations (region II), then the behavior is fluctuational and the indices are of Ising type. If there are still no developed fluctuations (region I) or they have been already suppressed by a field (region III), the behavior is mean-field type and the indices are classical. If the fluctuations are not sufficiently developed or due to a field have started to become anisotropic, then there is a transitional type of behavior, and the indices have intermediate values.

On this assumption the distinction of various nonideal-specific systems should manifest itself only in the change in the positions of the boundaries of region II between two crossovers up to its complete (in certain cases) disappearance. Although this model is quite sketchy it allows us to further elaborate the position of these boundaries.

Firstly, if the internal fields guarantee the suppression of fluctuations in the whole temperature range, right up to temperature which hardly differs from the critical one, then region II can be extremely narrow or even completely absent. Interestingly, in [70] the following was said concerning the critical behavior of superconductors: "... due to the region's narrowness, the effect (*heat capacity anomaly*, D.I.) will be smeared even if the sample deviates slightly from the ideal." If we take into account that the paper [70] was dealing with the narrowness of the fluctuation region, then this phrase can be considered as a logical justification of our proposed scheme. It later turned out that systems with long-range forces (superconductors, ferroelectrics, etc.) nearly always demonstrate the classical, mean-field behavior [24].

Secondly, the position of the boundaries of the fluctuation region should differ not just for various nonideal systems but also for different physical properties of the same system which behave anomalously in the vicinity of the critical point.

This clarification seems not only quite essential but also physically justified as, despite the general reason for anomalous behavior near the critical point, the anomalies of various physical quantities are described by critical indices of different values. Indeed, compressibility divergence does not need, in principle, to be as close to the critical point as, for example, the heat capacity anomaly of the same liquid under the same conditions. However, the Ginzburg criterion, which is based on the heat capacity anomaly, does not suggest this a priori [70]. This means that curves, similar to that shown in Fig. 2.1 for various system properties, might not coincide with each other and could be shifted by one to another along the horizontal axis. In other words, each specific physical property has its own disposition of regions I, II, and III (Fig. 2.1) relative to the critical point (see also Figs. 2.5, 2.6, 2.10, 2.12, and 2.16). As we shall see below, it is extremely important to take these circumstances into account when checking the universality relations which exist between the critical indices.

Finally, if it is correct to say that in the limit there are no other critical index values than Ising-type or classical ones, then even with a few disturbing factors the system can only once pass on from region II to III due to the effect of the strongest of them. Other "fields" should not seemingly influence the behavior of critical indices after such a transition. It is, of course, possible to combine the effect of different "fields," but in this case there will also be only one second crossover which can change its position on the temperature or density axes. It should also be mentioned that the curve in Fig. 2.1, modeling the critical behavior of the system in the presence of a field, can be naturally presented in a more complex way: the flat section does not have to reach the limiting value of the corresponding critical index, the "slopes" might not be smooth, but, for example, stepped, etc. All these, and other, questions can and should be answered by the microscopic theory of critical behavior of *real*, nonideal, systems a particular example of which we find in the already cited paper [78]. Now, in order to illustrate these ideas we consider the examples of the evolution of critical indices in the vicinity of the critical point for various real systems in the presence of different distorting factors.

2.2
Critical Indices: External Field Effects

2.2.1
Critical Index β

To determine the critical index β it is necessary to study the coexistence curve in liquids, and the magnetization curve for magnets. To see how the critical index β changes, data from the coexistence curve of pure SF_6 (see Figs. 1.4 and 1.5) are represented here in a log–log scale (Fig. 2.2). In this case the curve's asymmetry (see also Figs. 1.6 and 1.7) in the region near T_c ($|\tau| < 10^{-5}$) is manifested especially clearly.

Figure 2.3 clearly demonstrates the changes in the critical indices of the coexistence curve as the system approaches the critical point. However, if this behavior were previously noticed only in the papers [87, 89, 91, 126], it would have now received additional confirmation in the results of other research [98]. Apart from the basic feature which is the transition from "almost" Ising-type value of the index β "far" from the critical point to its classical value in its nearest

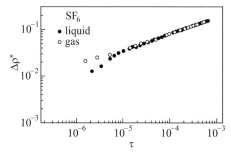

Figure 2.2 Dependence of $\Delta\rho^*$ on τ in a log–log scale [141]. Plots are made using the (*pVT*)-data from [87, 91].

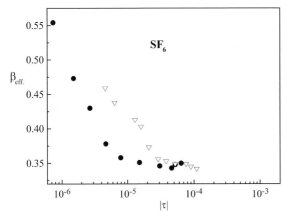

Figure 2.3 Dependence of the critical index β for the coexistence curve of pure SF$_6$ on the proximity to the critical point [2, 141]. Data refer to (•) liquid branch [87, 89, 91, 126] and (▽) values along the whole coexistence curve [98].

vicinity, this figure clearly shows that the growth of β when the vessel's height is 30 mm [98] starts at a greater distance from the critical point than in a vessel of 8 mm. The latter can certainly be taken as an additional argument for the idea that it is gravitation that is the cause of the observed changes of this index.

2.2.1.1 The Gravitational Effect

Assuming in the first approximation that the coexistence curve asymmetry is totally determined by gravitation let us consider how it effects the temperature of the appearance or disappearance of one of the matter phases in a piezometer on both sides from critical density. The method of fixing of the system's transition from a two-phase state to a single-phase one and vice versa does not seem to play an important role. It seems plausible to assume [87, 89] that gravitation, which compresses matter when there is an unlimited compressibility growth, encourages the formation of a liquid phase from a gas one and makes the reverse transition harder to perform. Continuing this thought, we can say that instead of point *A* in Fig. 2.4, point *A′* should be experimentally found and instead of point *B*, point *B′*. Thus, the coexistence curve becomes asymmetric (see the dotted line in Fig. 2.4). Despite the simplicity of this explanation, we believe that it illustrates quite accurately what is happening physically.

Remember also that according to the model of critical behavior of real liquids suggested above, fluctuations in this region are suppressed by external fields, in this case by gravitation. Consequently, concrete interactions, which previously did not play a significant role on the background of the growing correlation length, could now become very important. We do not mean here the "primitive" gravitational effect, which leads to a redistribution of density along the height of the vessel, as an additional pressure due to the column's height (the height of the piezometer

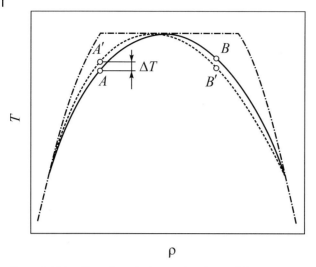

Figure 2.4 The effect of gravitation on the shape of the coexistence curve [87, 91].

was 8 mm) did not exceed 60 Pa, which is comparable to the measurement error and was only ∼0.0016% from critical pressure.

Another argument in favor of the suggested explanation could come from Frenkel's theory of heterophase fluctuations [167]. This theory's basic idea is that close to the transition points, from one phase to another, fluctuations of different signs are unequal with respect to the probability of occurrence and when averaged their influence is not destroyed, which, finally leads to the appearance of nuclei of the second phase within existing (stable) phase. It seems to us that the role of gravitation becomes even more natural. Indeed, Frenkel's mechanism relates to ordinary, noncritical fluctuations. It is suggested that it could be used effectively not only far from the critical point, where fluctuations are small, but also when critical fluctuations are suppressed by some external field such as, e.g., gravitation.

Under these conditions not only the Kadanoff–Wilson "interaction symmetry type" [26, 40, 41, 61, 166] but also the details of these interactions determine the critical behavior. Therefore, it seems to us that the above-mentioned considerations (Fig. 2.4) concerning the role of gravitation (and in fact of any other field) in the nearest vicinity of the critical point do not show ad hoc statements. Gravitation (or any other field), which suppresses fluctuations in the immediate vicinity of the critical point (region III, Fig. 2.1), removes all restrictions on the possibility of influence by the concrete details of interaction on the development of critical phenomena (see below for comparison with metals).

2.2.1.2 The Influence of Surface Forces

It is interesting to note that an analogous suggestion concerning the asymmetry of the coexistence curve to that proposed above was made later for binary mixtures placed between two parallel walls (see [168, 169] and references therein), and for pure liquids (SF$_6$) placed in nanometer pores [170], subjected to the action of surface tension forces. It turned out that if the surface forces favored one of the mixture's components, then the curve was shifted toward one (bigger or smaller) concentration, and if they favored another component then the shift was found to be in the opposite direction. In another paper [137] by the RG-method for a two-dimensional Ising model, a critical binary mixture placed in a narrow slit between two parallel plates was investigated. It was shown that if the gravitational field applied compensates the oppositely directed surface forces, then the system returns to a state of "ordinary" critical point behavior. This unequivocally, in our opinion, demonstrates the equivalence, according to the results, of different types of forces near the critical point on the critical behavior.

2.2.1.3 The Influence of Fields: Comparison with Magnetic Materials

The next two figures show the dependence of the critical index β on the degree of proximity to the critical point for both branches of the coexistence curve of SF$_6$ (Fig. 2.5) and for ferro- and antiferromagnets (Fig. 2.6) (the data for Fig. 2.6 and Table 2.1 are taken, with some abridgement, from Kadanoff et al. [48]). It seems to us that the qualitative and partly even quantitative similarity of these graphs is symbolic. The experiment thereby shows that both liquid and ferromagnets, on one hand, and gas and antiferromagnets, on other hand, behave in a similar way in the vicinity of the critical point. However, we are not talking about the well-known similarity between the critical behavior of liquids and magnets which results from the scaling hypothesis [61, 62]. In this case we are only talking about the way of evolution of the critical index β in both systems to its classical value [89, 91, 141] as they approach the critical point. Moreover, "far" from the critical point both liquid and magnets have the same index $\beta \approx 0.33 - 0.35$, close to the Ising value $\beta = 0.325$. Near the critical point the situation changes dramatically, β_{liq} and

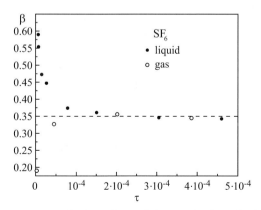

Figure 2.5 The dependence of the critical index β for the gas and liquid branches of the coexistence curve of SF$_6$ on its proximity to the critical point [141]. Curves are plotted using data from [87, 89, 91].

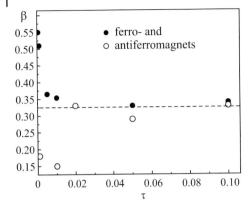

Figure 2.6 Dependence of the critical index β for ferro- and antiferromagnets on their proximity to the critical point [141]. Curves are plotted using data from Table 2.1 [48].

Table 2.1 Critical parameters for ferro- and antiferromagnetics [48].

	Ferromagnet			**Antiferromagnet**	
Substance	**Δ|τ|**	**β**	**Substance**	**Δ|τ|**	**β**
Fe	$2 \times 10^{-3} - 10^{-1}$	0.34 ± 0.02	MnF_2	$8 \times 10^{-5} -$ 2×10^{-2}	0.335 ± 0.01
Ni	$5 \times 10^{-4} - 10^{-2}$ $10^{-2} - 1.6 \times 10^{-1}$	0.51 ± 0.04 0.33 ± 0.03	$CuCl_2 \cdot 2H_2O$	$5 \times 10^{-4} -$ 10^{-2} $10^{-2} - 10^{-1}$	0.18 ± 0.07 0.29 ± 0.03
EuS	$10^{-2} - 10^{-1}$	0.33 ± 0.015	$KMnF_2$	$10^{-2} - 10^{-1}$	0.33
$YFeO_3$	$2 \times 10^{-4} - 3 \times 10^{-3}$ $10^{-2} - 3 \times 10^{-1}$	0.55 ± 0.04 0.354 ± 0.005	$CoCl_2 \cdot 6H_2O$	$10^{-2} - 10^{-1}$ $5 \times 10^{-2} -$ 2×10^{-1}	0.15 ± 0.02 0.23 ± 0.22

$\beta_f \Rightarrow 0.5$, while β_{vap} and $\beta_{af} \Rightarrow 0.25$ (for some, as yet unknown reasons, this trend for liquid and ferromagnets is more clearly expressed than for gas and antiferromagnets).

It is known that both adjusted values are classical in the sense that they are a consequence of the hypothesis concerning the possibility of the Taylor expansion of thermodynamic functions near the critical point. Meanwhile, the equation of state is analytical at the critical point (this equation could, for example, be the van der Waals equation), and if the first, not equal to zero, isothermal pressure derivative with respect to density is the third derivative, then $\beta = 0.5$, and if it is the fifth one then $\beta = 0.25$ (see, e.g., [26, 27]).

The mentioned analogy in the behavior of liquids and magnets (here and further we are only talking about the analogy in the behavior of the critical index β for both systems) can be, at least, qualitatively understood [89, 91] within the lattice-gas model, where phase space partition into cells itself is equivalent to an effective repulsion between particles (see, e.g., [171]). If most cells are not filled (low-density state, gas), then the repulsion prevails over attraction and the situation becomes similar to that for antiferromagnets. When the cells are mainly filled, there is an analogy between liquid and ferromagnets. We hope that a deeper theoretical foundation will be found for this nontrivial fact.

2.2.1.4 Comparison with Metals

The asymmetric shape of the coexistence curve was also obtained as a result of quite impressive experiments on liquid metals Rb and Cs [129] (Figs. 2.7 and 2.8). The authors of this experimental paper, as well as theorists discussing these results [122–124], were sure that a singular diameter of the coexistence curve in accordance with Eq. (1.7) was observed there. There is really no doubt about this. If the coexistence curve is asymmetric, then the diameter is curvilinear. The authors of these works suggested that the reason for this is that in metal liquids the interparticle potential depends more strongly on the nature of the

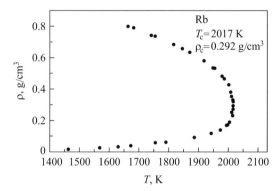

Figure 2.7 Coexistence curve for rubidium. Figure taken from [141] plotted there by using data from [129].

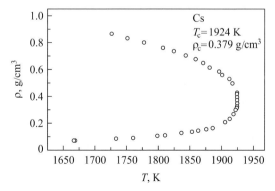

Figure 2.8 Coexistence curve for caesium. Plotted originally in [141] using data from [129].

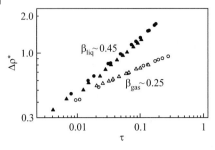

Figure 2.9 Coexistence curves for rubidium (•, ○) and caesium (▲, △). The filled symbols relate to the liquid branches and the empty ones to the gas branches. Plotted in [141] using data from [129].

electron gas, screening the Coulomb interaction, than on the atomic properties themselves [124].

The data of the coexistence curve for both metals analyzed in the same way as for SF_6 (see Fig. 2.2) are shown in Fig. 2.9. From this figure it is clear that both branches of the coexistence curve for both metals shown in a log–log scale are almost ideal straight lines. Moreover, the liquid branches of Rb and Cs have the critical index β for the whole range of temperatures close to 0.5 ($\sim$0.42), while for the gas branches it is close to 0.25. This is very similar to what was found for SF_6, as well as in the case of ferro- and antiferromagnets (see above). Similar results are also known for other metals (see, e.g., [172] and references therein). The difference from a simple liquid consists in the fact that for SF_6 only the top of the curve behaves in that way, while for metals the whole curve is asymmetric.

The question arises whether this similarity suggests that here we have also a situation where "near the phase transition point, independent of the details of interaction, external factors of different nature can influence in the same way on the type of behavior of such physical quantities" [89, 91]. Is it possible to explain the observed change in the critical index β to its classical value as resulting in particular from the electron gas in metals near the critical point where the effective long-range interaction arises and consequently classical critical indices appear? Clearly, for such a complex problem all these suggestions must be taken as hypothetical.

According to our proposed model (Fig. 2.1) this means that pure liquids and metals fall into two different regions of classical behavior. Nonpolar SF_6, which experiences the second crossover under the influence of an *external* (gravitational) field, appears in region III, while metallic liquid due to the *internal* (Coulomb) field does not completely leave region I. This can be indirectly seen in the different character of temperature dependences for a simple liquid (Figs. 2.3 and 2.5) and for metallic ones (Fig. 2.9).

The question why, as follows from the analysis, a value of β for the gas branch of a coexistence curve and antiferromagnet, on the one hand, and that for liquid branch and ferromagnet, on the other, are not the same, but exhibit different classical critical indexes, 0.25 and 0.5, respectively, remains open. Such an extremely interesting and undoubtedly fundamental question will need to be further explored in future.

Thus, despite the fact that the authors of [122–124, 129] believed that liquid metals are completely different as compared to ordinary liquids in their critical behavior, there is much evidence to state that they fit into the general picture. In the case of ordinary liquids, gravitation can suppress critical fluctuations only in the nearest vicinity of the critical point, and consequently, the second crossover appears. In that case when one or another internal field, such as, e.g., Coulomb field in metals (as a result of the possible influence of electron gas or some other mechanism) which initially guarantees long-range interaction, acts in a system then, the probability of its leaving region I (Fig. 2.1) turns out to be extremely low (see also [70]). However, all these ideas have probably only a heuristic value and as the authors of [124] correctly remarked, after they had tried to explain the effects observed in liquid metals in [129], "… it remains an important open problem to develop a microscopic theory of critical phenomena in systems with such rapidly changing electronic structure as metallic fluids."

Nevertheless, attempts to move ahead in this direction continue. Without going into details, as for us only the fact itself is important, it should be mentioned that the paper [173], for example, not only suggests a mechanism of deformation of fluctuations by electrical fields, in particular, in polymer solutions but also offers a more general idea that due to the influence of external fields on approaching the critical point "fluctuation behavior can once more become mean-field." These ideas, in turn, also agree very well with Beysens and colleagues's [174] assertion that "external fields or disturbances such as shear flow, turbulence, temperature gradients, gravity, walls influence (pores, surface forces) etc. can lead to anisotropy and even to destruction of the fluctuations" (compare with Fig. 2.15 taken from one of Beysen's earliest works [96]). It is quite remarkable that this statement appeared in the paper [174] where for the first time the morphology of critical fluctuations of the order parameter was directly analyzed using their visualization with subsequent computer analysis. It is quite clear that all these ideas are similar to those which were stated in one [87, 89, 91, 126] or another [141, 142] form starting from 1974. In recent years, as we can see, totally different experiments confirmed them.

2.2.2
Critical Index δ

2.2.2.1 The Influence of Gravitation
By analyzing small sections of the critical isotherm step by step, getting closer and closer to the critical point, it became possible to calculate the changes in the effective value of the critical index δ in its nearest vicinity [2, 141]. It was also found that it gradually changed toward its classical value $\delta = 3$ (see Fig. 2.10), which was explained by the influence of gravitation, as it had been for other indices [2, 90, 91, 141]. An analogous result was obtained and similarly interpreted in [98]. Figure 2.10 clearly demonstrates the compatibility of the behavior of the

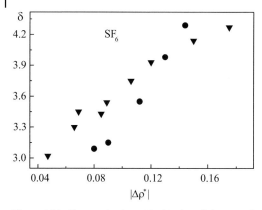

Figure 2.10 Changes in the critical index of the critical isotherm on approaching the critical point on density [2, 141]. (•) plotted using data from [88, 91]; height of the piezometer 8 mm; the data relate only to the "liquid" branch; (▼) plotted using data from [98]; height of the piezometer 30 mm; the data refer to both branches.

critical index δ in the nearest vicinity of the critical point with the suggested model (Fig. 2.1). This confirms thereby that the transition to classical indices can take place when real system "moves" toward the critical point not only on temperature but also on density.

2.2.2.2 The Influence of Coulomb Forces

Similar changes in the index δ were noted not only in response to gravitation but also under the influence of Coulomb forces [175]. The authors, from Kiev University, of this paper investigated the near-critical isotherm (the lower critical point) of the binary mixture 3-methylpyridine–heavy water using optical methods. It turned out that by adding just 0.3% of ions (Na^+ or Cl^-) the critical index δ changes from 4.4 to 3.05 ± 0.15 for both branches of the critical isotherm (Fig. 2.11). We can assume that in this case the weak Coulomb field, which guarantees its long-range action, transfers the system into the region analogous to region III and not region I (Fig. 2.1) as the system had already been in the fluctuation zone II before the field was introduced. We also suggest that if the experiment was carried out with different ion concentrations, then, with sufficient sensitivity, the whole "spectrum" of values of δ between 3.0 and 4.4 could be found.

This leads to the idea of using the Coulomb field to investigate the influence of different (in view of universality) fields as it, unlike gravitation, can be easily modified by changing, for example, the ion concentration. Doing so, however, instead of the critical isotherm, more convenient to study compressibility, isothermal (for pure liquids) or osmotic (for binary mixtures) as, in this case, the position of the second crossover will change on the temperature axis, which is more convenient than on the density one.

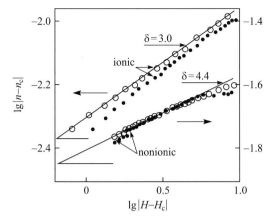

Figure 2.11 "Height" dependence of the difference of the refraction index Δn for the upper (•) and lower (○) branches
of the ionic and nonionic (without salt) isotherm for the binary mixture 3-methylpyridine–heavy water [175].

Such behavior, but for other indices, γ and ν, was found in the presence
of Coulomb forces for other binary mixtures (see, e.g., [176–178]), and also for
polymers (see, in particular, [173]). Without going into details the explanation of
the authors of [176–178] came down to the following. There are two mechanisms
of phase separation. The first is the so-called Coulomb immiscibility which is
determined by the long-range effect of electrostatic forces and the second the
so-called solvophobic immiscibility, which is related to the predominance of
short-range forces, such as hydrophobic interaction in aqueous solutions.

In the first case the long-range Coulomb forces could lead to the suppression of
fluctuations and consequently the system's behavior will become close to classical,
mean-field one, while in the second case they could lead to the Ising model. In
essence, this interpretation rather corresponds to that which was put forward above
for a similar occasion and also with the idea that weakly changing Coulomb forces
can also be responsible for establishing ferroelectric phase, as was mentioned
in [48]. It is also possible to explain the asymmetry of the coexistence curve,
supposing that the gravitational field encourages the formation of one phase and
prevents the appearance of another one (see Fig. 2.4 and comments to it in the
text).

2.2.3
Critical Index γ

2.2.3.1 The Influence of Gravitation
Isothermal compressibility (Eq. (1.12)), which describes the reaction of the system
on the external mechanical action, should, in essence, be more exposed to the
gravity influence. The (pVT)-experiment on pure liquid (SF_6 and CO_2) [89, 91, 99]
which we discussed in Chapter 1 (see Figs. 1.16–1.19 and comments in the text)

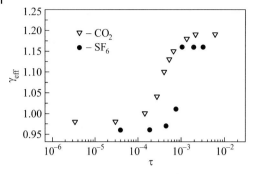

Figure 2.12 Changes in the critical index γ on approaching the critical point [2]. Plotted using data from [91, 92, 163] ($\bullet$), height of the piezometer 8 mm; $\triangledown$ [99], height of the piezometer 30 mm.

confirmed this. Analysis of all these data shows [2, 141] that the change in the critical index γ toward its classical value really happens on 1.5-2 orders further away from the critical point with respect to reduced temperature (Fig. 2.12) than for the analogous change in the critical index β [87] (Figs. 2.2, 2.3, and 2.5). It is clear from Fig. 2.12 that the difference in the positions of the second crossover in the experiments [91, 92, 163] on the one hand, and in [99] on the other, is in qualitative agreement with the dependence of the difference in densities of the investigated substances and in the heights of the used piezometers. This is the reason why gravitation can be considered as the main reason for this. It is also clear that the character of the change in the critical index γ in the region II$\Leftrightarrow$III follows the discussed model (Fig. 2.1) more fully than the other indices (β and δ).

We shall now briefly look at the most significant results for the gravitational effect obtained in theoretical papers. The first paper to look at is Hohenberg and Barmatz's paper [115] which used the so-called linear model of the parametric equation of state [179] to calculate the density distribution with respect to the height of the vessel, the isochoric heat capacity behavior, and the low-frequency sound velocity in the presence of gravitation. According to the authors, their main aim was to obtain corrections caused by gravitation and not to clarify the form of the exact equation of state [115]. It was in this paper that the formula was obtained for the dimensionless width (Δ^*) of the flat part of the top of the coexistence curve:

$$\Delta^* = \frac{\delta}{\delta + 1} \left(\frac{h_0^*}{D_0} \right)^{1/\delta}, \quad h_0^* \equiv \frac{\rho_c g H}{p_c}, \tag{2.1}$$

where D_0 and δ are the amplitude and the critical index of the critical isotherm (see Eq. (1.10)), respectively, and H is the vessel's height.

We used this relation to determine the critical index δ with the help of the experimentally measured width of the flat part of a coexistence curve. Figure 2.13 shows the dependence, satisfying Eq. (2.1), for different δ in application to the experiment on SF_6 (see Eq. (1.11), Figs. 1.5–1.7)

$$h_0^* = 3 \times 10^{-5}, \quad D = 1.70, \quad \Delta^* \approx 0.02. \tag{2.2}$$

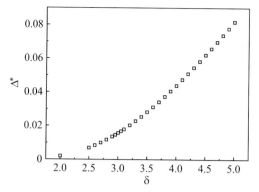

Figure 2.13 Relation between the critical index δ and the reduced width (Δ^*) of the flat part of the coexistence curve. Plotted in accordance with Eq. (2.1) [141].

From Fig. 2.13 it follows that for SF_6 in the presence of gravitation $\delta \approx 3.2$. This value also agrees well with its direct determination from experimental data in the region near the critical point on the critical isotherm itself (see Fig. 2.10). We should also add that although Eq. (2.1) was obtained using the "linear model" of the equation of state, the authors of [115] especially stressed that its application is of a general character and does not depend on a concrete model. The result shown above seems to confirm the validity of this opinion.

In [115] the value of the temperature range (τ) from where the influence of gravitation on compressibility turned out to be significant was also obtained as

$$\tau_2 = \left(\frac{\Gamma_0^+ h_0^*}{B_0} \right)^{1/(\beta\delta)} , \tag{2.3}$$

where B_0 and β are the amplitude and the critical index of the coexistence curve (see Eqs. (1.4) and (1.5)), and and Γ_0^+ is the amplitude of isothermal compressibility in the single-phase region (see Eq. (1.12)). Unlike [115] we used "2" as the subscript for τ [141], as according to the ideology of the model of critical behavior for real systems (see Section 2.1), it is just the second crossover [2] for compressibility that the temperature position is given, in our opinion, by Eq. (2.3). The evident nonuniversality of τ_2 should be noted (compare with the Ginzburg criterion [70]).

A paper close in spirit to [115] is [180], in which Chalyi and Chernenko studied in particular the relation between true and experimentally measured static critical indices under gravity with finite vertical dimensions of the vessel. It was shown that gravitation leads to the fact that the measured effective isothermal compressibility index can be less than the true one, $\gamma_{eff} < \gamma_{true}$, as was experimentally obtained in [89, 91, 92]. It should be recalled here that the central point of the approach taken in [115] was the idea that a fluid near the critical point is like a local homogeneous system, despite the influence of gravitation. However, as the authors of several papers on this topic mention (see, e.g., [181–183] and references therein), "in the immediate vicinity of the critical point, as a consequence of the interaction between adjacent layers with different densities, this suggestion becomes invalid." In other words, "gravitation changes the local properties of liquid, modifying, thereby, the very nature of the phase transition near the critical point." Further in these papers it was noted that the development of a consistent theory of the "intrinsic" gravitational effect would require a RG-analysis. However, as such an analysis was not available at the time (as far as we are aware, it is still not available), the van der Waals "square-gradient" theory (see, e.g., [184]) was used as an alternative, although it has many disadvantages [181].

It should be mentioned that similar ideas first appeared still in 1979 in the paper of Moldover et al. [185]. Moreover, the idea of the universality of the influence of different factors on the system's critical behavior was stated, in a general way, earlier in our papers [89, 91]. The main conclusion that the authors of the works quoted above [181–183, 185] came to is that gravitation modifies the system's properties near the critical point, and correlation functions become anisotropic. Due to the latter a difference between correlation lengths which are parallel and perpendicular to the gravitational field direction should be made.

Critical Fluids under Shear Flow: In principle, the result just mentioned is analogous to that obtained previously by Onuki and Kawasaki in their RG-description of critical systems under shear flow [78]. They showed that in this case an anisotropy of critical fluctuations arises and, consequently, the transition of critical indices from Ising-type to classical values takes place. Simultaneously, and independently, Beysens and colleagues found the same behavior of the susceptibility for the binary mixture aniline–cyclohexane [95] and then for other binary systems [96, 97] under shear flow.

Susceptibility, which determines the system's reaction to the most varied external influences, is a more general thermodynamic characteristic of a system than isothermal compressibility. However, the critical index participating in the description of both characteristics of a system, as is known, is the same index γ. In Fig. 2.14, Beysens and colleagues's experimental results are shown. Figure 2.15 represents a "naive," according to Beysens [96], illustration of the shear influence on the shape of the critical fluctuations. It is clear that in the direction, perpendicular to the flow, the fluctuations appear very noticeably deformed.

If we return to the RG-analysis of critical behavior in flow [78], we should emphasize that its authors concluded that shear flow, which suppresses long wave fluctuations, leads to the following consequences:

- the phase transition takes on a mean-field character;
- the spatial correlation function becomes anisotropic and stretched along the direction of the flow;
- the critical temperature decreases;
- the equation of state and critical indices become mean-field type.

From all the analysis carried out (see also Fig. 3 on page 481 of the paper [78]) it is clear that we talk here about the transition of the system to classical behavior

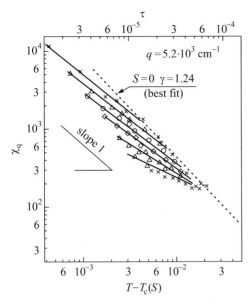

Figure 2.14 The susceptibility dependence of the binary mixture aniline–cyclohexane on temperature and shear [95,96].

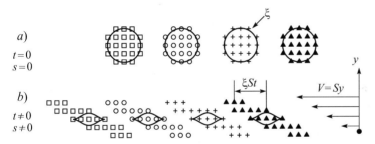

Figure 2.15 "Naive" picture of the effect of shear on the critical fluctuations [96]: (a) without shear, (b) with shear.

specifically in the nearest vicinity of the critical point, i.e., from region II to region III in the above-discussed model (Fig. 2.1). In other words, this work first described theoretically, for the case of shear flow, the second crossover. Not long before this, it was experimentally found on pure liquid (SF_6) in the gravitational field [87].

2.2.4
Critical Index α

2.2.4.1 The Influence of Gravitation
As has been already mentioned in Chapter 1, all the known experimental attempts to use an equation such as Eq. (1.14) to determine the value of the critical index α from the $(p - T)$-dependence along the critical isochore [91,132,152] were unsuccessful,

despite the extreme accuracy of the measurements. It was not possible to determine the critical index of isochoric heat capacity better than by an order of magnitude and the most likely value turned out to be the classical value $\alpha = 0$ rather than the one for the Ising model, $\alpha = 0.11$. To find out why such serious efforts led to such uncertain results, we once more turn to the influence of gravitation[2] [91, 126]. As is known, gravitation distorts the heat capacity anomaly (see, e.g., [40]), making it like a jump for which $\alpha = 0$.

It is usually necessary to spend a great deal of efforts to discover the true picture of the critical behavior of heat capacity under conditions of real experiment for pure fluids. It is typical that A. Voronel who, with colleagues, eventually discovered experimentally the singular character of the heat capacity behavior [80], in one of their first papers on this topic obtained exactly a jump [186]. To be fair, it should also be noted that heat capacity singularity was found long before this by Skripov and Semenchenko on a system (binary mixture) [79] which was almost free from the influence of gravitation. Unfortunately, this paper received attention only about 15 years later [187].

One could imagine that it would be a good idea to carry out such experiment during a space flight where it should be free from the distortions caused by the Earth's gravitational field. However, this is also not that simple. German researchers led by J. Straub spent more than 10 years studying the isochoric heat capacity anomaly under conditions of microgravity (see, e.g., [188–190]). Although this topic lies outside this book's scope, it is interesting to look at some observations from these works. It turned out that the experiments performed during the first German Spacelab Mission D1 (1985) have revealed that "the measured c_v-values were even more strongly distorted than those obtained with the same facility and the same ramp rates under $1\,g$ conditions" [190]. A few years after the experiment in space, the authors wrote, "... it became evident that the flattening of the c_v-curves observed could be caused by a nonuniform mass distribution and the slow mass diffusion process" [189, 190] (for further results concerning the features of the experiment under microgravity conditions see also [191] and references therein). Thus, by getting rid of the influence of gravitation, the scientists, who sent their apparatus for studying critical phenomena into space, acquired other, no less serious problems. In the absence of gravitation the slowing down of heat and mass transfer and also the influence of surface forces, which as we saw acted just like gravitation, became predominant factors.

Returning to the result $\alpha \approx 0$, obtained from the $(p - T)$-dependence analysis along the critical isochore, we can suggest that this value is also a consequence of the distortion of the heat capacity anomaly by the Earth's field, or in the spirit of this book, this is the critical index of isochoric heat capacity in the presence of gravity [2, 126, 141, 142].

---

2) An alternative explanation is that neither p nor μ has any peculiarities on the critical isochore. It is interesting that the "rectilinear diameter" is automatically fulfilled.

2.2.5
Critical Index of the Correlation Radius ν

The critical index of the correlation radius, ν, due to the methods of its determination, undoubtedly refers to the dynamics of critical phenomena (see Chapters 4–7). However, it is responsible for the behavior of such an important physical quantity for critical phenomena, as a whole, that it exceeds the limits of whatever their division.

There are not so many works devoted to the determination of the index ν, which are of interest in the light of all what was said above. One analysis, which fulfills all requirements, is the research carried out by Fabelinsky and colleagues [192]. In this analysis, the temperature dependence of the correlation radius of concentration fluctuations in the region of the double-critical point of a guaiacol–glycerine solution was studied [192]. This system is interesting as being a homogeneous one at all temperatures and concentrations. However, with the addition of a small amount of water, it starts to form a closed region, in which the solution decomposes into two phases. Moreover, in temperature–concentration coordinates this region becomes bigger the more water is added [192]. The viscosity of this system near the double-critical point (the occurrence of such special point is the distinguishing feature of this and similar systems), i.e., near the state where the upper and lower lines of the critical points meet, was investigated using both a viscometer and correlation spectroscopy in the single scattering mode (the basics of correlation spectroscopy and its development for multiple scattering systems, and also for systems near a critical point, are discussed in Chapters 4–6). This approach made it possible to determine the temperature dependence of the critical index ν. As the authors themselves remarked, it was surprising that on approaching the double-critical point the index ν changed from the value close to an Ising type to one which was typical for mean-field behavior.

Figure 2.16 shows the changes in the critical index of the correlation radius. It is clear that this dependence does not differ fundamentally from those obtained for the coexistence curves (Figs. 2.3 and 2.5) and magnetization (Fig. 2.6), on the critical isotherm (Fig. 2.10) and at studying of isothermal compressibility in the single-phase region (Fig. 2.12). With a slightly greater scatter of points, compared to the (pVT)-method, the tendency for the critical index ν to change to its classical value can also be quite clearly traced. The observed evolution of the critical index of the correlation radius allows one to suppose that a similar behavior is not limited only by the set of static critical indices, but is a property of the nearest vicinity of the critical point for real systems [2, 141]. Without suggesting any reason for such behavior of the critical index ν the authors of [192] noted that "binary solutions with small additions of a third component ... are items where small external actions lead to serious intermolecular effects." In the spirit of all the above said, as it seems to us, this statement gives the explanation that is the essence of the problem: near the critical point, small, in the usual sense, disturbances (whether they are gravitation, internal or external electric fields, shear, etc.) always lead to serious consequences. Thus, it seems that we are dealing with another system where the second crossover, i.e., a passage from region II to region III was found (Fig. 2.1).

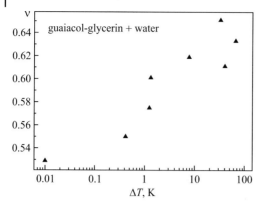

Figure 2.16 Dependence of the critical index of the correlation radius on the proximity to the double-critical point on temperature in the presence of water. Plotted in [2, 141] using data from [192].

This statement is supported by the fact that on approaching the double-critical point the critical index ν changes from Ising-type to classical values but not in the reverse direction (Fig. 2.16). The exact reason for this type of behavior in the particular system under investigation is unknown (the general features of systems with a double-critical point are discussed, for example, in [193]). In an attempt to understand a mechanism of such "uncommon" behavior of the critical index ν the authors of [192] tried to apply the Ginzburg criterion. However, as is known, to really use this criterion is not so simple: the data required for its calculation are not available, as a rule [70]. They were also not available for this particular system [192]. But even if this attempt were successful, then the position of the first crossover would have been determined by the Ginzburg criterion, whereas in this work it was the second crossover that was probably first established for the critical index ν.

Consequently, the question of applying the Ginzburg criterion is not really that simple. This is so not only and not mainly because it is sometimes difficult to obtain the necessary data, as it turned out to be the case in the work of Fabelinsky and colleagues [192]. There is, in our opinion, a more serious deep reason for the difficulties connected with its applicability. In [70], where it was first obtained, a nonmonotony of the dependence of its behavior on the degree of proximity of the system to the critical point was not supposed to occur. The appearance in the immediate vicinity of the critical point of the second crossover, the passage from fluctuation-type to classical behavior of matter, indicates that near the critical point there are two changes in the mode of critical indices behavior, as shown in Fig. 2.1, instead of only one change. This circumstance, in turn, requires, in our opinion, a more complex dependence of a criterion like the Ginzburg criterion on temperature (density).

The Ginzburg criterion, in the form as it was obtained many years ago [70], which determined the region where the classical theory can be applied, gives the position for only one crossover, in accordance to our terminology, for the first crossover (a lot of attention is now devoted to this crossover due to its practical implications for constructing equations of state valid in wide parameter ranges (see, e.g., [194] and references therein)). This attitude can be understood easily as in 1960, when this paper [70] was written, no date were available pointing out the need to account for the additional peculiarities in the behavior of critical indices in the nearest vicinity of the critical point, which were only found later. Nevertheless, it is not so much the consequence of the Ginzburg criterion itself as due to the ideas on which it was based [70], that it seems natural that in the nearest vicinity of the critical point of a real, nonideal, system the influence of different disturbing factors can lead to mean-field behavior appearing once more.

The classical values ν and γ were also obtained by Skripov and Kolpakov [82] in the research carried out near the critical point of SF_6 and CO_2 by static light scattering measuring its integral intensity (for more on this method and its possibilities, see Chapter 7). In this scrupulous, still prelaser, work the features of light scattering near the critical point of sulfur hexafluoride and carbon dioxide were first studied with ingenious thermal pressure adjustment. The setup's geometry was such that passing and scattered light beams lay in the vertical plane. Moreover, an incident light had to go up a significant distance (21 mm) through the scattering medium, crossing sections with density difference caused by gravitation. It is possible that with such geometry the optical experiment can be considered, in a certain sense, as equivalent to a (pVT)-experiment, which always deals with average density. Seemingly, this is what explains why classical values of critical indices were obtained in [82]. It should be mentioned that the authors did not calculate the critical indices themselves. However, it was done much later in [195] where their data were employed.

2.2.6
Micellar Systems

Micellar solutions, which are binary mixtures of water and nonionic surfactants (here we mean compounds like $CH_3(CH_2)_{i-1}O(CH_2CH_2O)_jH$, which are abbreviated here to C_iE_j), have become very widely used for studying critical phenomena only in the last 20 years (see, e.g., [196–199]). In pioneering investigations carried out by Italian physicists [196, 197] it was shown that the phenomena in these systems, which had been previously taken as being a result of the growth of accidentally forming aggregates, were actually a true phase transition for a binary mixture, corresponding to a decomposition into two phases, a phase enriched by a surfactant and a phase depleted with respect to it. What is more, in one case, such a system exhibited fluctuational behavior while in other cases classical behavior of the indices γ and ν was found. This result received a great attention from theorists (see, e.g., [200–202]) and also from experimentalists. Dietler and Cannell [198] repeated the experiments made in [196, 197] on the same substance, $C_{12}E_8$, produced by the same company, Nikko Chemicals Co., but unlike [196, 197] they obtained Ising-type and not classical values. The most remarkable situation, however, occurred when they repeated the experiment using an "Italian" sample instead of "their own" $C_{12}E_8$ and obtained the same results as in [196, 197], i.e., the critical indices once more demonstrated classical behavior. At the end of their article, Dietler and Cannell put forward the obvious suggestion that the result was somehow related to the kind of sample employed.

Actually, it is well known that the properties of systems near the critical point (in micellar solutions this is the CMC, the critical micelle concentration) can change dramatically due to the presence of minor traces of impurities (see, e.g., [203]). However, in none of the cited papers [196–198] the purity of the sample was especially checked (for example, by the concentration dependence of the surface tension, a method that is quite simple and effective [203]). It is therefore quite possible to assume that an unnoticed or accidentally introduced impurity in the "Italian" sample could have played a role similar to the small addition of water in the case of the guaiacol–glycerine binary mixture [192]. Of course, it is quite possible that there are other reasons, as a surfactant–water mixture is quite a complex system. However, we are here mainly interested in the fact that this system, although for an unspecified reason, repeatedly exhibited (as was independently confirmed twice) the second crossover in its critical behavior. If a concrete reason were discovered, then by varying the "strength" of its influence it would be possible to obtain on the same sample not only the crossover itself but also the whole spectrum of intermediate values of critical indices γ and ν, as it happened in [95–97] when investigating the temperature dependence of susceptibility in flow.

There is another paper [199] which is remarkable in the considered context as here in the analysis of micellar solutions like $C_{12}E_5$ + water an unusual value for the coexistence curve critical index $\beta = 0.25$ was found. This was the same critical index as obtained in the immediate vicinity of the critical point for the gas branch of the coexistence curve of SF_6 (Fig. 2.5), Rb and Cs (Fig. 2.9), and also for the antiferromagnetic magnetization curve (Fig. 2.6). In addition, the coexistence curve's diameter was rectilinear, while the critical indices γ and ν were also close to their classical values. Subsequently, however, Ising-type critical indices were obtained in the same laboratory on the same system [204]. In this connection, it is unclear which of these papers is close to the truth as in the second work [204] the results of the first investigation [199] were not discussed at all. It is quite possible that, as in the case of $C_{12}E_8$, both sets of critical indices are valid for the same reasons as discussed above.

2.2.7
Influence of Boundaries: Finite-Size Effects

The quite comprehensive and independent part of the theory of critical phenomena, the so-called *finite-size scaling* (see, e.g., [54, 204–211]) is remarkable, in particular, due to the fact that it is, strangely enough, quite close to the description of experiments under real conditions of the nearest vicinity of the critical point. Although critical phenomena are theoretically characterized by the singularity of the correlation length, in fact this singularity, as all the authors of the cited works mention, is limited by various external perturbations, impurities, different external fields, etc. The presence of such disturbances leads to a *rounding* and/or a shift of the critical point [54, 205]. Besides, a large number of physical systems (such as magnets and binary systems with large, but finite interaction radii, polymer mixtures with large, but finite size of the chains [208, 211]) begin to exhibit mean-field critical behavior [209] since, in accordance with Ginzburg's ideas and the Ginzburg criterion [70], critical fluctuations cease to play a decisive role under these conditions.

As one example of such kind of behavior similar to the one described above for liquids and ferromagnets, we would like to discuss briefly the behavior of the critical index β, as it was obtained comparatively recently by a Monte Carlo experiment on a two-dimensional model employing a truncated Lennard–Jones potential and taking into account size effects [210]. For the coexistence curve, in a system where linear dimension was the control parameter, the crossover from Ising-type behavior (for a two-dimensional model $\beta = 1/8$) far from the critical point to the mean-field ($\beta = 1/2$) behavior close to it was found. In accordance with the concept proposed by us this effect has to be interpreted as the second crossover. Unfortunately, it was not possible to obtain such a crossover in this work for a three-dimensional model. The author himself writes that dimensional effects are also present here, but for a Lennard–Jones-type potential these effects, as a rule, are not that clear.

In addition, it is well known that, in order to go over from performing the calculations on two-dimensional to three-dimensional models, a large increase of computational power is required. For example, to analyze the behavior of a polymer chain near the critical point on a three-dimensional lattice using the Monte Carlo method it took 3000 (!) hours of CRAY-supercomputer calculations [211].

2.2.8
Results and Consequences

Finally, it should be mentioned that in each of the investigations examined here, concrete systems not related to each other, and correspondingly different physical fields which have an influence on the character of their critical behavior, were studied. It is indicative that in the whole history of research on critical phenomena there have been no more than 20 such investigations out of the tens of thousands of works on critical phenomena. These papers, as they dealt with different systems and fields, do not seem, at a first glance, to have anything in common (the authors hardly ever referred to each other's works as they presumably considered their scientific areas to be very different).

However, if one considers them together, then the results of these investigations manifest itself in a completely new way. Now they make it possible to bring to light new features of critical behavior of real, nonideal, systems. In particular, it is possible, in accordance with the proposed phenomenological model of the critical behavior of such systems (see Section 2.1) and taking into account all the examined experimental information, to formulate some theorem-like statement "on even crossover number": a real system, in contrast to an ideal one, demonstrates near the critical point either two crossovers or their number is zero [141, 165].

If in the system a certain internal field (such as a Coulomb field) acts from the very beginning and this field does not allow critical fluctuations to develop, then the system, as it approaches the critical point, does not perform any crossover and demonstrates along the whole "path" the mean-field, classical behavior. In this case, region II (Fig. 2.1) just does not appear. For example, both ferroelectrics and superconductors behave in this manner. If such a field is not acting in the system from the beginning, then it first passes from region I to region II (first crossover) and then, by necessity, some influence is found which at first deforms and then suppresses at all the critical fluctuations. Therefore, there will be a passage in the opposite direction (second crossover) and the real, nonideal, system once more will begin to demonstrate the mean-field, classical, behavior.

Unlike for the phenomenological approach, a microscopic approach to describing critical behavior of real systems does not yet exist. In the special case of systems under shear flow, for which RG-analysis was carried out [78] (this analysis remained for a long time the only one), the obtained results agreed well, as we have already mentioned, both with the proposed model and with the results of other systems with different disturbing fields.

Another successful attempt to construct a RG-theory. of critical behavior for diluted solutions of electrolytes was undertaken quite recently [212]. The aim of this paper was to describe critical behavior of weak electrolytes in the first crossover region for subsequent comparison with the results of a special experiment on light scattering in a ternary system 3-methylpyridine + water + sodium bromide [213]. It turned out that the theoretically and experimentally obtained results for the critical index γ agreed qualitatively with each other [212, 213]. The comparison of Figs. 2.1 and 2.17 shows, in turn, that in the first crossover region, for which the curve was calculated (Fig. 2.17), its shape fully corresponds to the character of the curve, modeling the critical behavior of real systems according to Fig. 2.1.

It is important to underline that, from Fig. 2.17, for large values of the dimensionless constant b, which reflects the interaction between the order parameter and the density of ions, the "critical index vs. temperature" dependence has been obtained not with one, but, actually, with two crossovers (the curve corresponding to $b = 3$ in Fig. 2.17) [212]. Although the author did not comment on this fact, it seems to us that it, together with the results from [78], can serve as an additional experimental confirmation, within the limits of the RG-approach, of the phenomenological model examined here. Also, it is clear from Fig. 2.17 that by varying the concentration of the added ions, i.e., regulating the interaction constant, it is possible, as we suggested above, to obtain the whole spectrum of values from fluctuation to classical type for one or another critical index.

Experimental investigations of different systems show, as we could confirm in the above discussions, that the restoration of the mean-field critical behavior in the nearest vicinity of the critical point is not a special singular event but rather the manifestation of a universal property of real, nonideal systems. In fact, all examined and similar (see, e.g., [196, 197, 210]) systems clearly demonstrate in their

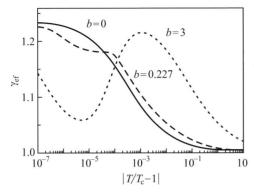

Figure 2.17 Temperature dependence of the effective critical index of susceptibility γ_{eff} of the model systems for various values of the dimensionless constant b, which characterizes the interaction between the order parameter and density of ions [212]. The curve, where $b = 0$, corresponds to a non-ionic solution.

critical behavior a second crossover induced by different reasons. In liquid–vapor systems [87, 89, 91, 98, 99] it is the intrinsic gravitational effect, creating an effective long range action through the interaction between macroscopic layers of matter that starts to play a role [181]). In binary mixtures it is a shear flow [78, 95–97]. In a double-critical point case [192] some other disturbing factors (e.g. the Coulomb field of water dipoles) operate. It is the variety of concrete disturbing factors which lead to the same effect, the restoration through suppression of large-scale fluctuations of mean-field behavior, that convinces us of the universality of such a picture for real systems.

Is such behavior in the nearest vicinity of the critical point really universal? Keeping in mind Feynman's words (see introduction and [3]) let us be careful. The final, conclusive answer to this fundamental question will only be obtained from a consistent microscopic theory of critical behavior of nonideal systems.

2.2.9
Some Unresolved Problems

Above, considering the state of affairs with respect to the behavior of matter in the nearest vicinity of a critical point, in particular, the experimental data on the second crossover [2, 87, 141, 164, 165], we already could get convinced that a variety of unresolved problems still exist in critical phenomena physics. Here we would like to address in more detail some of such problems further. But here we will be concerned mainly with details of the first crossover.

The first crossover entered into our model by a logically consistent way (as discussed in Section 2.1), but without any discussion about its experimental manifestations. Note that Ginzburg's paper [70] was written in 1960 when the modern theory of phase transitions did not exist yet. The purpose of this paper was to show that the nature of all phase transitions, both of the second order and the first order close to the second, was the same, and distinctions had only a quantitative character. The basic thesis in this analysis was: "all transitions of second order are the same and differences between them are connected only with relative weight of bulk and correlation energies."

The Ginzburg criterion, based on such a concept, actually assumes that the dependence of all properties on fluctuations near a critical point has an universal character, i.e., the first crossover position depends only on the type of the system under consideration and it should be the same for various thermodynamic quantities.

This idea has found its confirmation in some theoretical investigations [214] (see also [215, 216] and references therein). In these works, for three-dimensional Ising-like systems the temperature dependences of the critical indices α, β, and γ were determined in the critical region. It is possible to get an impression on their form by having a look at the basic curve (solid line) in Fig. 2.17. These dependences appeared similar in shape for all three indices and also they were absolutely equally located along the temperature axis. However, the latter property does not seem, in our opinion, to be physically justified [217]. Indeed, the existing distinction in critical index values between "big" indices, such as γ (index of compressibility or isobaric heat capacity), and "small" ones, such as α (index of the isochoric heat capacity), indicates convincingly that the basic mechanism (in particular, fluctuational) has different levels of influence on one or the other thermodynamic properties near a critical point [217].

Experiments in the second crossover region, as we have seen, essentially confirm this idea: for the "big" index $\gamma \approx 1.2$ (a continuous growth of compressibility is directly connected with a growth of fluctuations [166]) considerable changes begin practically at two decades on temperature earlier, than for a relatively "small" index $\beta \approx 0.35$ (Figs. 2.3, 2.12, and 2.26). In this connection, a similar question concerning the first crossover emerges: in which manner and where does it actually manifest itself experimentally with respect to various thermodynamic properties of real systems in a critical region?

For convenience of the subsequent comparison we recall here once again a summary of the fundamentals of the modern theory of critical phenomena (see, e.g., [26, 40, 41, 52]). It is widely accepted (see, e.g., [215, 216]) that "critical-point universality originates from the long-range nature of the order-parameter fluctuations"[3]. In the so-called critical region fluctuations are so large that microscopic details of short-range intermolecular interactions become insignificant. In this region, the fluctuational (Ising-like) behavior should be observed. In practice, one can consider $\tau \approx 10^{-2}$ or a few kelvin around an ordinary critical temperature as the critical region [215]. Outside the interval $\tau > 10^{-2}$ one should expect a

---

3) Actually, this statement is not absolutely true, since all versions of the mean-field theory which do not consider fluctuations at all also predict a universal critical behavior for very different physical objects, such as fluids, magnets, and alloys. Therefore, here it would be more exact to speak not about the "universality," in general, but about the universality of a certain type.

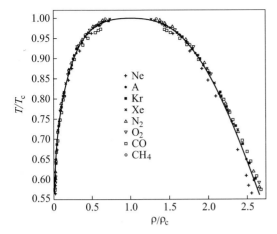

Figure 2.18 The generalized coexistence curve of eight pure liquids from the original Guggenheim's work [42].

crossover to classical, mean-field (van der Waals-like) behavior (first crossover). The experimental facts accumulated in the world literature on critical phenomena can be used both for checking of these concepts and for tackling the question of the first crossover formulated above.

In the work "The principle of corresponding states" [42], which has already become classical, Guggenheim has set as the purpose to find out the degree of applicability of this principle to various thermodynamic characteristics of matter in different states of aggregation. In particular, he plotted the generalized coexistence curve of eight pure liquids (see Fig. 2.18), and draw the attention to the fact that it does not agree with the van der Waals equation, not being of parabola type but, more likely, of a cubic shape. Since then this result was included into many publications on critical phenomena, as an example that the critical index β is equal to 1/3, instead of 1/2. However, neither Guggenheim himself to whom this result was only a byproduct of his analysis, nor, unfortunately, those who subsequently quoted his work, have noted another surprising fact: the generalized coexistence curve can be well approximated by a simple power dependence with an exponent $\beta = 1/3$ extending to states very far from the critical point, from an almost zero value of the reduced density (ρ/ρ_c) up to values $(\rho/\rho_c > 2.5!$ The reduced temperature range also was incredibly wide, $0.55 \leq (T/T_c) \leq 1$ (Fig. 2.18).

In fact, it is such a behavior that proves the validity of the "principle of corresponding states." Note that this "principle," apparently, relies on the existence of a particular point with special status, the critical point, distinguished by nature itself. The fact that the "principle" is valid on such wide temperature intervals implies that a system 'feels' his own critical point even very far from it. The behavior of the thermal conductivity critical enhancement which is also observed for several tens of degrees apart from a critical point (see also Chapter 7 and Fig. 7.3) testifies this statement as well. The nature of such effects and feasible mechanisms of their realization are not clear. Moreover, such a situation, we should repeat, seems rather surprising from the point of view of the widely accepted today critical phenomena concepts.

In recent years, $(p\rho T)$-experiments of highest accuracy have been performed on a set of pure substances in a wide range of state parameters, states in close vicinity of the critical point including [218–224]. Figures 2.19 and 2.20 show the generalized curves for various pure substances plotted after proper mathematical processing of initial data from these works in the form of log–log graphs [225]. As it is known, such a type of representation allows one not only to determine effective critical indices, but it is a highly sensitive test of the form of the crossover behavior as well.

One can see from Figs. 2.19 and 2.20 that the "principle of corresponding states" holds true both for the coexistence curves and for the compressibility. Attention is attracted by the fact that, in wide temperature intervals, experimental data are well approximated by one-term formulas (as a calculation shows, the error is within the limits of ∼1%). This circumstance, in turn, confirms once again the conclusion that an application of polynomial formulas for data presentation in the vicinity

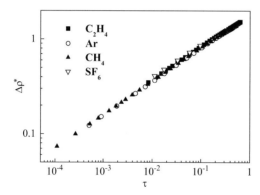

Figure 2.19 Generalized coexistence curves for ethylene [218], argon [219], methane [220], and sulfur hexafluoride [221]: $B_0 = (1.814 \pm 0.008)$, $\beta = (0.3498 \pm 0.0014)$.

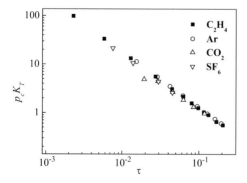

Figure 2.20 Generalized compressibility for ethylene [222], argon [223], methane [224], and sulfur hexafluoride [221]: $\Gamma_0^+ = (0.087 \pm 0.004)$, $\gamma = (1.144 \pm 0.013)$.

of a critical point, most likely, is superfluous [141, 153, 225, 226]. Note that the opportunity of the description of thermodynamic characteristics of pure liquids in such a wide temperature interval by such a simple equation with such high accuracy is of huge interest for the practical use of $(p\rho T)$-data. The most surprising fact is, however, that in the whole investigated region (Figs. 2.19 and 2.20), far beyond the limits of "a few kelvin around a critical temperature" any "kinks" which would indicate a change of values of critical indices β and γ, and, hence, also a change of the character of critical behavior are not observed (cf. with the second crossover region, Figs. 2.2, 2.3, and 2.12).

The effective critical indices obtained in these cases have the values $\beta = (0.3498 \pm 0.0014)$ and $\gamma = (1.144 \pm 0.013)$ which are different both from classical and from Ising ones. At the same time, as the "left" border of the temperature range on each of these graphs practically coincides with the beginning of a passage into the second crossover area (see, e.g., Figs. 2.3 and 2.12), it is possible to believe that for real pure liquids an Ising-like region with its values of critical indices, if it exists, is extremely narrow [225].

As all these important conclusions are a consequence of generalized dependence analysis they demand a more detailed consideration. Therefore we plotted and analyzed also similar dependences, but now, for individual substances. In Fig. 2.21 the dependence $\Delta\rho^* = f(\tau)$ is plotted according to initial data from [45]. In this paper, by the way, the fact that $\beta \neq 1/2$ has been observed for the first time. In this work Verschaffelt, apparently, also for the first time has applied a log–log scale analysis. He found that for pure CO_2 the critical index has the value $\beta = 0.3434$ and keeps this value constant up to the end of the investigated temperature interval, down to a temperature more than $60\,^\circ$C below the critical one (Fig. 2.21). Our additional computation shows that for CO_2: $\beta = (0.3434 \pm 0.0006)$ and $B_0 = (1.698 \pm 0.004)$. Similar to the case of the generalized dependences (Figs. 2.19 and 2.20), the presented data are well approximated by smooth straight lines and do not demand for their description with an accuracy of ~1% a more complex dependence than a one-term formula. And this result was obtained as far ago as in the year 1900! Then, after the corresponding mathematical processing

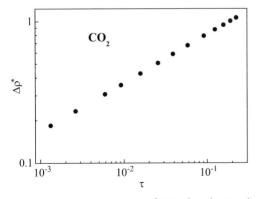

Figure 2.21 Coexistence curve of CO_2 plotted using the experimental data from [45].

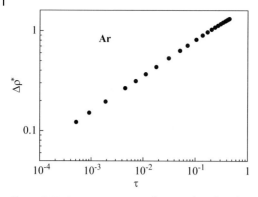

Figure 2.22 Coexistence curve of argon plotted in the range $\Delta T \equiv |T - T_c| \approx 60$ K and $\Delta \rho^* \geq 1.3$ using experimental data from [219]: $B_0 = (1.785 \pm 0.004)$, $\beta = (0.3519 \pm 0.0011)$. The linear correlation coefficient R is equal to $R = 0.9999$.

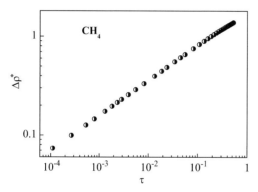

Figure 2.23 Coexistence curve of methane plotted in the range $\Delta T \equiv |T - T_c| \approx 100$ K and $\Delta \rho^* \geq 1.3$ using the experimental data from [220]: $B_0 = (1.782 \pm 0.003)$, $\beta = (0.3493 \pm 0.0009)$, $R = 0.9999$.

of initial data from modern ($p\rho T$)-analyses, we plotted the dependences for the coexistence curve (Figs. 2.22 and 2.23) and for the isothermal compressibility along the critical isochore in the single-phase area (Figs. 2.24 and 2.25) for some pure substances of various chemical structures. It is evident that the coexistence curves of argon and methane in a log–log scale are ideal straight lines with an accuracy not worse than 0.5% and correspond to the one-term equation $\Delta \rho^* = B_0 \, |\tau|^{-\beta}$ (Eq. (1.5)) up to the triple point (!). As to the isothermal compressibility, the dependence $p_c K_T = \Gamma_0^+ \tau^{-\gamma}$, plotted in the same way, is linear with high accuracy as well. Results for other investigated substances (see, e.g., [218, 221, 223, 224] and references therein) do not exhibit, as our analysis shows, any distinctions from the results presented here.

As a whole, the absence of visible "kinks" on these dependences means that in a real experiment on pure liquids it is not possible, strange as it may seem, to

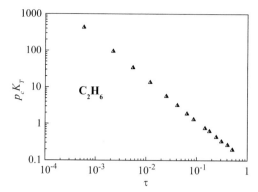

Figure 2.24 Compressibility of ethane plotted in the range $\Delta T \equiv (T - T_c) \approx 150$ K using the experimental data from [227, 228]: $\Gamma_0^+ = (0.088 \pm 0.002)$, $\gamma = (1.140 \pm 0.005)$, $R = 0.9999$.

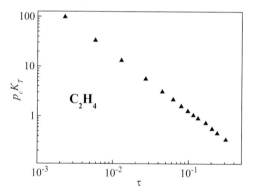

Figure 2.25 Compressibility of ethylene plotted in the range $\Delta T \equiv (T - T_c) \approx 90$ K using the experimental data from [222, 228]: $\Gamma_0^+ = (0.087 \pm 0.001)$, $\gamma = (1.154 \pm 0.004)$, $R = 0.9999$.

find out any change of the mode with a passage to the mean-field behavior on any distance from a critical point. This newly discovered situation, certainly, requires a theoretical explanation. First of all, it will be necessary to specify the role and influence of fluctuations on the nature of critical phenomena. Melville S. Green, the chairman of the organizing committee of the already multiply mentioned famous conference, still wrote [229]: "Why do deviations from the classical theory of critical opalescence occur within hundredths of a degree of the critical point while deviations in thermodynamic quantities show nonclassical behavior much farther away?" Despite the very high intensity of research on critical phenomena any answer to this question has not been obtained up to now. From the above discussion together with the fact that the nearest vicinity of a critical point of nonideal systems is a van der Waals-like region (the second crossover) it follows most likely that a certain correction of the modern theory of critical phenomena is required, at least, in application to pure liquids.

2.3
Critical Indices and Amplitudes

As was already mentioned, critical indices play a key role in the theory and practice of critical phenomena, as one or another set of their numerical values satisfies one or other critical behavior model (see, e.g., Table 2.2). It is no less important that the thermodynamic and statistical fundamentals in the assumption of the validity of scaling lead to the existence of a well-defined set of universal relations, independent of the model, between critical indices (see, e.g., Eqs. (2.4)–(2.11)). Although these some combinations of critical amplitudes appear to be universal (see, e.g., [76]), on their own, unlike indices, they are not universal. The numerical values of these relations are also unequivocally connected with the universality class of the investigated system, which makes it possible to use them for the determination of critical behavior and universality classes just as effectively as critical indices. Therefore, it is just as important to carefully determine them experimentally as it is for critical indices.

Furthermore, an analysis of the wide base of experimental thermodynamic data near a critical point showed [141, 226, 237] that there should exist definite correlations between the values of critical amplitudes and indices, related to the critical behavior of the same physical property. The search for a correlation between the values which characterize the different aspects of behavior and properties of matter makes up the base of the methodology of thermodynamic similarity (see, e.g., [238, 532]), including the similarity hypothesis (scaling) widely applied in the theory of critical phenomena. However, for the correlations which we shall be dealing with here, the theory of critical phenomena has still not been examined. All these questions will be discussed below.

2.3.1
Universal Relations Between Critical Indices

As was discussed in Chapter 1, the only strict relations established for critical indices are the following inequalities:

Table 2.2 Example of sets of values of critical indices.

Model	Critical indices				
	α	β	γ	δ	ν
Mean field	0	0.5	1.0	3.0	0.5
Ising ($d = 3$, $n = 1$)	0.110	0.326	1.239	4.80	0.630

$$\gamma \geq \beta(\delta - 1), \tag{2.4}$$

$$\alpha + \beta(\delta + 1) \geq 2, \tag{2.5}$$

$$d(\delta - 1) \geq (2 - \eta)(\delta + 1), \tag{2.6}$$

$$\nu(2 - \eta) \geq \gamma, \tag{2.7}$$

$$\alpha + 2\beta + \gamma \geq 2, \tag{2.8}$$

$$\gamma(\delta + 1) \geq (2 - \alpha)(\delta - 1), \tag{2.9}$$

$$d\nu \geq 2 - \alpha, \tag{2.10}$$

$$\nu(d - 1) \geq \overline{\mu}, \tag{2.11}$$

where η is the Fisher critical index. In order to prove these relations, the definitions of the corresponding thermodynamic quantities, and rigorous physical statements such as, for example, the nonnegativity of heat capacity, and simple geometric considerations like convexity or concavity of thermodynamic potentials as functions of their variables (see, e.g., [26]), are usually employed (for more details on the relation between geometry and thermodynamics, see Chapter 3). However, the simplicity of the preconditions does not imply that the evidences are simple.

Within the limits of the scaling hypothesis each of these inequalities transforms to equality. This makes it, in principle, possible to check its validity experimentally. Obviously, it is possible at least to an extent in which the conditions of a real experiment manage to differ an equality from an inequality. On the other hand, if the similarity hypothesis itself does not raise any doubts and can be considered valid, then these relations make it possible, determining experimentally the values of only two critical indices, to calculate the values of the others.

Before moving on to such a calculation and discussion of concrete experimental results which refer to critical indices and the relations between them, we shall discuss the significant limits put on this process by the properties of the model of critical behavior (see Section 2.1). Let us look at Fig. 2.26 where the evolution curves for the critical indices β (Fig. 2.3) and γ (Fig. 2.12) are shown together for the sake of convenience. This figure directly shows one of the above-discussed features of the critical behavior of real systems in the nearest vicinity of a critical point: *the second crossover region for different physical properties of the same system under comparable conditions is observed at a different distance from a critical point.* From the curves shown (Fig. 2.26) it follows that the second crossover for the "strongest" critical index of isothermal compressibility (γ) is far from the critical point almost by a factor of 10^2 with respect to τ, than the similar transition in the case of a relatively "weak" coexistence curve index (β). While checking the feasibility of the universal relations between critical indices it is extremely important to take this fact into account.

It is now clear that for *real* systems it is necessary to extract effective values of all critical indices, which can take part in such a check, not from the same

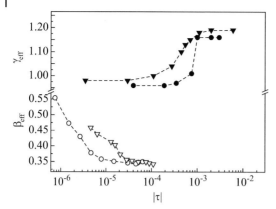

Figure 2.26 Effective critical indices of the coexistence curve β_{eff} and isothermal compressibility γ_{eff} for pure SF_6 ($\circ$ [87], $\bullet$ [89, 163], $\triangledown$ [98]) and for CO_2 ($\blacktriangledown$ [99]) in the second crossover region. Data for β and γ refer to the ranges $\tau < 0$ and $\tau > 0$, respectively. The piezometer height was equal to 8 mm [87, 89, 163] and 30 mm [98, 99]. For convenience, experimental points are connected by dotted lines.

temperature (density) range, as it is usually done, but from the corresponding regions of the curve (Figs. 2.1 and 2.26) for each index separately. Moreover, it is also important that all necessary critical indices (at least three) are determined on the same experimental setup and on the same sample. Existing experimental data confirm this conclusion even though, for understandable reasons, they are not enough numerous. Therefore, here we are using the results of our own precise $(p\rho T)$-experiment on SF_6 [89, 91, 141], in the course of which the critical index values β, γ and δ and their amplitudes (B_0, Γ_0^+, and D_0) were obtained (in the presence of gravitation) from both sides from the position of the second crossover for each index (see Eq. (2.12)):

$$T_c = (318.7355 \pm 0.0002)\,\text{K},$$

$$\rho_c = (738.83 \pm 0.15)\,\text{kg m}^{-3}, \quad \rho_c' = (741.60 \pm 0.15)\,\text{kg m}^{-3},$$

$$P_c = (38.3281 \pm 0.0003)\,\text{atm} = (3.759987 \pm 0.00003) \times 10^6\,\text{Pa},$$

$$|\tau| \geq 3 \times 10^{-5}, \quad B_0 = 2.054 \pm 0.021, \quad \beta = 0.3508 \pm 0.0013, \quad B_1 = 0,$$

$$|\tau| < 1 \times 10^{-5}, \quad \beta_+ \approx 0.5 \text{ (liquid)}, \quad \beta_- \approx 0.25 \text{ (vapor)}, \tag{2.12}$$

$$\tau \geq 1 \times 10^{-3}, \quad \Gamma_0^+ = 0.081 \pm 0.001, \quad \gamma = 1.16 \pm 0.03, \quad \Gamma_1 = 0,$$

$$\tau < 1 \times 10^{-3}, \quad \gamma \approx 1,$$

$$|\Delta\rho^*| \geq 0.08, \quad D_0 = 1.70 \pm 0.01, \quad D_1^+ = 1.40 \pm 0.05,$$

$$D_1^- = 0.06 \pm 0.02, \quad \delta = 4.30 \pm 0.01, \quad \Delta_+ = \Delta_- = \Delta = 0.49 \pm 0.02,$$

$$|\Delta\rho^*| < 0.08, \quad \delta \approx 3.2.$$

Thus, according to Eq. (2.12), one of the most important relations $\gamma = \beta(\delta - 1)$, which connects static critical indices, is fulfilled almost exactly both for the fluctuation region ($|\tau| \geq 3 \times 10^{-5}$ for β and $\tau \geq 1 \times 10^{-3}$ for γ), where $\beta = 0.3508 \pm 0.0013$, $\gamma = 1.16 \pm 0.03$, and $\delta = 4.30 \pm 0.01$, and for the region where fluctuations are already suppressed: $\beta \approx 0.5$, $\gamma \approx 1$ and $\delta \approx 3$. At the same time it is seen (Fig. 2.26) that the use of critical index values obtained in real experimental conditions (in the presence of gravitation in the case of SF_6) from the same intermediate temperature region for both indices, e.g., from $3 \times 10^{-5} < |\tau| < 3 \times 10^{-4}$, would appear wrong.

In turn, for this aim the necessity to use critical index values from different temperature regions could only mean that it is only in these temperature ranges the same critical behavior of those physical properties of a real system which are described by these indices can be observed. So, as we are talking about gravitation it follows that its influence on compressibility begins to display at a distance of 1.5–2 orders of magnitude far from the critical point on τ than on the coexistence curve. This seems perfectly natural from a physical point of view.

It also follows from Eq. (2.12) that the obtained critical indices, each in their own region II (Fig. 2.1), turned out to be not strictly fluctuational but only close to being so. To be more precise they have intermediate values between Ising type and classical one. This might point to the fact that either they are indirectly influenced by gravitation as well or the Ising model is not completely suitable for describing pure liquids (see also Section 2.3.3). As the universal relations were fulfilled it could be pointed out that this influence, which was observed at different "distances" from the critical point, was of the same type for all indices.

This result also confirms what was previously said that the positions of the region II boundaries (Fig. 2.1) should depend on the concrete physical property which demonstrates an anomalous behavior near the critical point. This conclusion is also confirmed, in particular, by the Monte Carlo experiments on a polymer system given in [211]. In this paper the authors varied the number of chain links, and thereby practically changing an effective long-range action, managed to create either classical or fluctuation type of behavior in the system.

Moreover, it turned out that for different system properties (order parameter and susceptibility were examined) the boundary between regions with different critical behavior types was found at different distances from the critical point. Analogous results were also obtained in [78], where for the same properties the change of indices (β and γ) under the influence of shear flow also took place at different distances from the critical point.

Now let us return again to discussing the results of our experiment on pure SF_6. The self-consistency of the numerical values of static critical indices within the framework of this experiment (Eq. (2.12)) makes it possible to use the remaining relations (Eqs. (2.5)–(2.11)) as equalities in order to calculate the substance's other critical indices. The critical index of isochoric heat capacity α, determined from different relations by taking into account Eq. (2.12), takes on values between 0.137 and 0.141. Its mean value $\alpha = 0.139$ agrees quite well for such a small index value with theoretical calculations: $\alpha = 0.07$ (ε-expansion), $\alpha = 0.125$ (series summation), $\alpha = 0.110$ (numerical solution of RG-equations (see, e.g., [40])), and also with experimental values which range from $\alpha = 0.04$ to $\alpha = 0.159$ (see, e.g., [52]). Considering how much the scaling hypothesis, according to S. Ma, is "simple and rough" [52] such a coincidence can be considered to be quite good.

Having α, it is possible to look at the determination of ν and $\bar{\mu}$. The coincidence of the obtained critical indices $\nu = 0.62$ and $\bar{\mu} = 1.24$ with values found in the literature can also be considered as quite good. For example, in study [230] for critical index of the surface tension which is rare in occurence the value $\bar{\mu} = 1.28 \pm 0.02$ for SF_6 was obtained. It should be mentioned that the authors of this paper, just as we did, carried out an analysis determining the critical index α from different relations between indices, and found for it the values which were located between $\alpha = 0.011$ from Eq. (2.10) and $\alpha = 0.109$ from Eq. (2.6). This comparison clearly shows that even the presence of good experimental data does not still guarantee automatically their self-consistence.

It is worth looking at the calculated value of the index of the anomalous dimensionality η, or the Fisher index. According to our data for SF_6 (Eq. (2.12)) its value changes within very narrow limits from 0.130 to 0.132. However, it is already remarkable that by such a calculation method it did not "dissolve" in the error of determination of other "large" indices, as its theoretical value is only $\eta = 0.031$ (see, e.g., [40]).

2.3.2
Universal Relations Between Critical Amplitudes

Modern theory of critical phenomena not only puts restrictions on the values of the critical indices (see, e.g., [26, 40, 52]) but also establishes universal relations between nonuniversal, in principle, amplitudes which take part in the description of equilibrium and kinetic properties of systems near phase transition points [76, 231–234]. However, unlike critical indices, for which all the known relations (except for those with space dimensionality) are satisfied for all the known models, including the "mean-field" model, the same combinations of critical amplitudes for various models have different values. Taking this fact into consideration, let us look at a situation typical only for the fluctuation region. But to examine this we shall use only some of similar universal relations:

$$R_\chi^+ = \Gamma_0^+ D_0 B_0^{\delta-1}, \tag{2.13}$$

$$R_\xi^* \equiv R_\xi^+ \left(R_\Gamma^+\right)^{-1/3} = \xi_0 \left(\frac{p_c B_0^2}{k_B T_c \Gamma_0^+}\right)^{1/3}. \tag{2.14}$$

Here ξ_0 is the correlation radius amplitude (direct correlation length), k_B is the Boltzmann constant, and R_χ^+, R_ξ^+, R_Γ^+, and R_ξ^* are the critical amplitudes.

A more informative relation, but at the same time not so commonly studied one, is the universal relation (2.13). To employ this relation in computations, it is necessary to experimentally investigate the coexistence curve (B_0), a sufficient number (no less than at least 10 to 20) of near-critical isotherms in the single-phase region (Γ_0^+), and the critical isotherm (D_0 and δ). Obviously, it is necessary to carry out all the measurements on the same experimental setup and on the same sample. The RG-theory predicts for this universal relation the result $R_\chi^+ = 1.60$, while the high temperature expansion method (HTS) gives $R_\chi^+ = 1.75$ [76].

For pure liquids, apart from SF_6, we are aware of only two papers [112, 136] where the experiment, using the dielectric method, was carried out in a sufficient full manner to calculate this universal relation. The data obtained in [112] made it possible to obtain $R_\chi^+ = 1.82 \pm 0.05$ for O_2. In [136], $R_\chi^+ = 2.05 \pm 0.8$ and $R_\chi^+ = 1.71 \pm 0.5$ were obtained for Ne and N_2. In the case of binary mixtures, which are in principle much easier than pure liquids to investigate near the critical point, this amplitude combination is only known for the mixture nitrobenzol–n-hexane ($R_\chi^+ = 1.75 \pm 0.30$) [94]. This is despite the fact that Beysens and colleagues studied the remaining universal relations between amplitudes for tens of other binary mixtures [93, 94, 235].

The substitution of the necessary values from Eq. (2.12) into Eq. (2.13) for SF_6 leads to the value $R_\chi^+ = 1.48 \pm 0.05$. As we can see this agrees well enough with both theoretical values and with the amplitude combinations obtained experimentally for other systems and with other disturbing factors. It is, therefore, possible to conclude that the universal relations between amplitudes are, less than critical indices and critical amplitudes themselves, subjected to the influence of such kind of factors. As the behavior of the universal relations between critical indices is the same it is possible for them to come to a similar conclusion as well. Below (see Section 2.3.3), while looking at the question of correlations between critical amplitudes and indices, we shall once more return to this amplitude combination.

We are not dealing here with the results, which were obtained using a quite widespread method when data from various sources are gathered to compare with theory (see, e.g., [49, 236]). It is impossible to call such a method of comparing experiment with theory in the critical area too successful. It is interesting that the authors of [236] write at the end of their article, although concerning some other topic: "as a final minor point we note that the experience we have had in 'forcing' data to fit the current theoretical framework involving correction terms has made it clear to us that almost any reasonable data set can be brought into very good agreement with theory, because of the tremendous flexibility afforded by the correction terms" [236]. I quote this in full so as to emphasize the fact that such a method of comparing experiment with theory is not useful to one or the other, and also because, unfortunately, it is in fact very often used. It should also be mentioned that the very idea of using additional, although incomplete information, for this purpose is extremely productive. However, the method how this idea can be realized, in our opinion, should be significantly different (see Section 2.3.3).

As for Eq. (2.14), it, like other universal relations between critical amplitudes, can be used to calculate those parameters which are part of it, but for whom no data were found in the examined experiment on SF_6. Thus, for example, together with Eq. (2.12) this relation makes it possible to find one of the fundamental characteristics of matter near the critical point, the direct correlation length ξ_0. Substituting into Eq. (2.14) the well-known theoretical values $R_\xi^+ = 0.26$ and

0.27 for the HTS-method and ε-expansion (order ε^2), respectively [232, 233] and $R_\Gamma^+ = 0.059$ and 0.066 for the HTS-method and ε-expansion, respectively [76], we obtain $R_\xi^* = 0.668$ for both calculation methods. Then, for SF$_6$, taking into account Eq. (2.12), we obtain $\xi_0 = 1.98 \times 10^{-10}$ m. In the previously mentioned paper [230], for SF$_6$, $\xi_0 = (2.0 \pm 0.3) \times 10^{-10}$ m was obtained, while in [135] $\xi_0 = 2.016 \times 10^{-10}$ m was found. It is clear that all the presented values of ξ_0 agree very well with each other.

We have earlier established that critical indices, which take part in the description of one or another physical property of a nonideal system, can have intermediate between Ising-type and classical effective values in quite a wide, and for each index its own, temperature (density) region (transition II$\Leftrightarrow$III, Fig. 2.1). These facts show that not only the effective values of critical indices themselves but also the corresponding amplitudes may satisfy the existing universal relations only if the correct temperature (density) regions of their experimental determination are chosen. Therefore, in the nearest vicinity of the critical point these relations are, in fact, only satisfied if for each concrete critical index and amplitude their values are taken from any, although identically positioned in relation to the second crossover, temperature (density) region (see Figs. 2.1, 2.3, 2.12, and 2.26). It is also necessary to account for the fact that, as will be shown below, there exist definite correlations between critical indices and amplitudes [226, 237].

2.3.3
Correlation Between Critical Index and Critical Amplitude Values

The idea that there should exist quite definite correlations between values of critical amplitudes and critical indices, corresponding to the critical behavior of the same physical property, has been proposed earlier [141]. An additional extensive analysis of experimental thermodynamic data near the critical point, carried out in recent years, shows that these correlations not only exist but can also be used very effectively [226, 237].

Correlation dependences between the amplitude–index pairs can be represented as simple power laws of the following form:

Coexistence curve:

$$B_0 \cdot \beta^{-b} = b_0, \tag{2.15}$$

Isothermal compressibility along the critical isochore in a single-phase region:

$$\Gamma_0^+ \cdot \gamma^g = g_0, \tag{2.16}$$

Critical isotherm:

$$D_0 \cdot \delta^{-d} = d_0. \tag{2.17}$$

In [226, 237], the existence of such correlations was confirmed by the results of the analysis of the whole set of available thermodynamic data for classical (nonquantum) pure liquids. In addition, a method of a proper test of the universality of various critical amplitude combinations was also proposed using results not from separate experiments but based on the similar generalized correlations.

In fact, if experimental thermodynamic data are used to determine the exponents b, g, and d and the "universal" constants b_0, g_0, and d_0 in Eqs. (2.15)–(2.17), then there arises a real possibility of properly checking the universality of different critical amplitude combinations using the results from all this information and not from separate experiments. In particular, the test of relation (2.13) can be performed if, using Eqs. (2.15)–(2.17), it will be given in the following way [226, 237]:

$$R_\chi^+ = \Gamma_0^+ D_0 B_0^{\delta-1} = g_0 \gamma^{-g} \cdot d_0 \delta^d \cdot (b_0 \beta^b)^{\delta-1}. \tag{2.18}$$

It should be emphasized that a positive result for such a test of relation (2.18) would also mean an adequacy of the approach, expressed by Eqs. (2.15)–(2.17). The set of critical indices β, γ, and δ, used for substitution in Eq. (2.18), should, obviously, satisfy the equation (2.4) considered as an equality.

It is suggested that all available experimental data should be used for such an analysis. This makes things significantly easier, as otherwise it would be necessary, for the independent determination of the universal constant value of R_χ^+ on the basis of one separate experiment, to investigate simultaneously the behavior of all necessary thermodynamic properties. Such complex precise investigations in the critical region are, as we have already seen, extremely difficult and are found therefore not so often.

In order to carry out such an analysis, precise experimental data were selected from the original papers [45, 87, 88, 99, 136, 154, 160, 163, 218, 220–223, 227, 239, 518] and from some reviews [39, 40, 49, 76]. These investigations cover a 100-year period out of the almost 200-year history of critical phenomena research. They start with Verschaffelt's, unfortunately, little-known work in which he has obtained almost the modern critical index values at the beginning of the 20th century in the famous Kamerlingh Onnes's laboratory in Leiden: $\beta \sim 0.34$ for the coexistence curve and $\delta \sim 4.26$ for the critical isotherm, by studying the critical behavior of CO_2 [45]. And they end with ($p\rho T$)-experiments carried out in recent years for a large group of ultrapure substances in Wagner's laboratory at Bochum University (Germany) [218, 220–223, 227, 239].

In certain cases the values of the critical amplitudes (B_0, Γ_0^+, and D_0) and indices (β, γ, and δ), which were necessary to carry out such an analysis, were contained in the works themselves (for example, in [40, 49, 76, 136, 154]). In other cases, to obtain the pairs of values $B_0 - \beta$, $\Gamma_0^+ - \gamma$ and $D_0 - \delta$ it was necessary to carry out a preliminary mathematical treatment of the original experimental data. For this purpose, all the experimental data, found in each work, were broken into small parts which were located at different distances from the critical point. Then for each of them, by using the least-squares method the values for the amplitude–index pairs were obtained. It was in that way the data from [160, 163, 218, 220–223, 239, 518] were approximated by Eqs. (2.15)–(2.17).

Then, using the *totality* of these data the amplitude–index dependences were plotted, as can be seen in Figs. 2.27, 2.28, and 2.29. As a result, on each graph the basic trend of the sought correlation was formed, confirming thereby the fact of its existence. The points which turned out to be positioned randomly relatively to the curve formed in this way were removed from the examination. This procedure is quite natural as it allows one to distinguish self-consistent experimental data from the rest. We shall look at the details of this procedure below.

The coexistence curve for pure liquids can be represented by one of Eqs. (1.4) and (1.5). The first of these relations, Eq. (1.4), is more traditional and supposes that critical behavior of real systems corresponds to the idealized model for which the equality $\beta = \beta_+ = \beta_-$ is typical. The other two representations, Eq. (1.5), are useful in that they make it possible to judge about possible asymmetry of the coexistence curve, which is displayed in the inequality between the critical indices β_+ and β_-. It is this type of analysis that made it possible to find the difference in the behavior of both branches of the coexistence curve of SF_6 in the nearest vicinity of its critical point (see Chapter 1, [87]). Later, this was used as the basis for the conclusion on the second crossover appearance, the transition in the nearest vicinity of the critical point of pure liquids from the Ising type of a behavior corresponding to the mean-field one [2, 164, 165].

To establish the constant numerical values in Eq. (2.15) the coexistence curves of different pure liquids were analyzed in the wide neighborhood of the critical point, with the obvious exception of the second crossover region. Moreover, curves with single-term expressions, Eq. (1.5), as well as more complex dependences taking into account the following terms in the expansion such as Eq. (1.4) were examined.

Figure 2.27 shows, in a log–log scale, the correlation between the critical index and amplitude for the coexistence curve of various substances, obtained as a result of the analysis. The line which is set out using "experimental" data of 22 pairs of values B_0 and β (Fig. 2.27) satisfies the linear dependence $\log B_0 = a + b \log \beta$ with $a = 2.25 \pm 0.10$ and $b = 4.298 \pm 0.217$. Moreover, the linear correlation coefficient is $R = 0.9754$, the standard deviation (SD) is equal to $SD = 0.014$, and the probability P of the absence of the correlation is $P < 0.0001$. After reducing this dependence to Eq. (2.15), the following result was finally obtained:

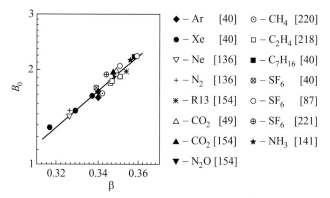

Figure legend:
◆ – Ar [40]   ⊙ – CH_4 [220]
● – Xe [40]   □ – C_2H_4 [218]
▽ – Ne [136]   ■ – C_7H_{16} [40]
+ – N_2 [136]   ⊗ – SF_6 [40]
∗ – R13 [154]   ○ – SF_6 [87]
△ – CO_2 [49]   ⊕ – SF_6 [221]
▲ – CO_2 [154]   ★ – NH_3 [141]
▼ – N_2O [154]

Figure 2.27 Correlation between critical amplitude (B_0) and critical index (β) for the coexistence curve of pure liquids [226].

$$B_0 \cdot \beta^{-b} = b_0 \Rightarrow B_0 \cdot \beta^{-4.30\pm0.22} = 178 \pm 40. \tag{2.19}$$

As can be seen, the largest deviation from the line (2.19) is found for Ar (the lower of the two points representing argon on the graph) and SF_6 from the paper [221]. However, if the corresponding values B_0 and β are substituted into Eq. (2.19), then for the constant b_0 we get $b_0 = 168$ for Ar and $b_0 = 190$ for SF_6. The error does not exceed practically the limits ±15, which is slightly less than 10% of the value of B_0.

It should be noted that, taking into consideration other sets of experimental data, the constants b and b_0, as well as their error, can change. However, as an analysis has shown, "correct," self-consistent, combinations B_0 and β will be located, with all changes, in the immediate vicinity of the approximating line. Hence, this fact makes it possible to consider the obtained dependence as a tool of testing experimental data with respect to their self-consistency.

As for the data scatter relative to the linear correlation, firstly, it seems to be unconnected to the difference in dipole moments of the substances analyzed. The polar N_2O (dipole moment 0.2 D), trifluorochlormethane (0.5 D), and ammonia, whose dipole moment is more significant (1.5 D) as well as all other substances, which have a zero dipole moment, are almost evenly distributed relative to the discussed line. Also this scatter could not be caused by the fact that one or other type of formulae (Eqs. (1.4) or (1.5)) of the representation (single or polynomial) along the whole coexistence curve or its separate branches were used by the authors in order to find the leading asymptotic term, as the corresponding data on the graph do not show any special regularity.

This result once more confirms that the application of polynomial formulas for an analysis of the data related to critical region is rather superfluous [141, 153, 226]. The presence of systematic errors in various authors' experiments should probably be considered as the main reason for the observed scatter, while in their own scales their results are usually much more precise (this is quite typical for critical phenomena research [39, 40, 141]).

On the other hand, the possibility should not be totally excluded of a certain difference on the background of the universality of critical behavior of separate groups of substances. In particular, quantum liquids helium-3 and helium-4 form such a group. As analysis showed [240], in this case, the correlation $B_0 \cdot \beta^{-b} = b_0$ between the experimental values B_0 and β in a log–log scale can also be approximated by a straight line. However, this straight line does not coincide with that for classical liquids but is located parallel to it. It should be emphasized that the obtained values of the exponent of b and the "universal" constant b_0 for helium also did not show any dependence either on its mass number or on the kind of formula, Eq. (1.4) or (1.5), for the coexistence curve representation used by the authors in the original works to analyze their data.

Isothermal compressibility in single-phase regions along the critical isochore in the asymptotic vicinity of the critical point can be exhibited using the well-known equation in dimensionless form, Eq. (1.12). Unlike the coexistence curve, for which there is a very wide array of experimental data, relating to pure liquids, analogous sets for compressibility and for the critical isotherm are noticeably more modest.

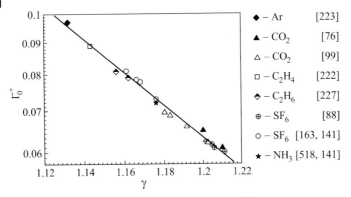

Figure 2.28 Correlation between critical amplitude (Γ_0^+) and index (γ) for the isothermal compressibility in the single-phase region [237].

Nevertheless, they were sufficient and made it possible to uniquely determine the necessary parameters in the dependences, Eqs. (2.16) and (2.17), in their analysis.

The correlation between the amplitude (Γ_0^+) and the index (γ) for the isothermal compressibility along the critical isochore in the single-phase region near the critical point for some pure substances is shown in Fig. 2.28 in a log–log scale, obtained using the described method. On this graph, the absence of points corresponding to the value $\gamma = 1.24$, which is well known for a three-dimensional Ising model result achieved with the ε-expansion, is noteworthy. At the same time, the paper [76] supplies lower values for this index: $\gamma = 1.20$ ($\Gamma_0^+ = 0.065$) and $\gamma = 1.21$ ($\Gamma_0^+ = 0.061$), obtained as a result of applying different theoretical models to describe experimental data in the vicinity of the critical point of CO_2, which quite well fit into the general picture (Fig. 2.28).

As for the pairs of values for Xe cited in [76]: $\gamma = 1.24$ ($\Gamma_0^+ = 0.065$), $\gamma = 1.21$ ($\Gamma_0^+ = 0.074$) and $\gamma = 1.18$ ($\Gamma_0^+ = 0.12$), we can see that they do not lie on the same line either with each other or with the other points on the graph (Fig. 2.28). Unfortunately, in most investigations of the compressibility of Xe, published later than the work [76], self-consistent pairs of the values $\gamma - \Gamma_0^+$ were not found as well.

The situation concerning SF_6 turned out to be very interesting [163]. It was assumed for a long time [141] that the pair of consistent values $\gamma = 1.16$ ($\Gamma_0^+ = 0.093$) corresponded to the experimental data obtained in [163]. Such consistency was confirmed by the practical coincidence of the value $R_\chi^+ = 1.70 \pm 0.06$, found for SF_6 using Eq. (2.13), with both its theoretical values [76]. However, it turned out that the point with the coordinates (1.16, 0.093) does not satisfy the obtained general regularity and is located noticeably higher in Fig. 2.28. As there was a trust in the quality of the experimental data [163] as well as in the validity of the idea developed here concerning the correlation between critical amplitudes and indices, it became necessary to reanalyze the results from [163]. In this work, 19 supercritical isotherms in the range $10^{-5} \leq \tau \leq 10^{-3}$ for the "far" critical region, adjoining $\tau \leq 10^{-3}$ (see Table 2 in [163]) were investigated. As a result, the critical index $\gamma = 1.16$ was obtained, while the amplitude (Γ_0^+) was not determined. A new analysis of these data showed [237] that the corresponding values of pairs amplitude–index, depending on the choice of the τ approximation range (excluding the second crossover region, where $\gamma \sim 1$, see Fig. 1.17) are 0.081–1.161; 0.0785–1.166; 0.0778–1.168; 0.073–1.176. These values are presented in Fig. 2.28. This example shows that the proposed correlations also have a significant prognostic potential. This once more confirms the possibility of their using to test experimental data on self-consistency.

It is worth noting that for NH_3, unlike all other substances (see Fig. 2.28), thermal conductivity and light scattering experiments instead of direct $(p\rho T)$-measurements were used. The measurements of the thermal conductivity coefficient of ammonia in a large temperature and pressure range including the critical region were carried out in [518]. The analysis of these data together with results of the light scattering intensity made it possible to obtain the following consistent values for ammonia: $\gamma = 1.176$ and $\Gamma_0^+ = 0.072$ [141] (see Chapter 7 for more).

As a result, for the "experimental" values of 19 pairs Γ_0^+ and γ the straight line was found (Fig. 2.28). It satisfies the linear dependence with a correlation coefficient $R = -0.996$, with a standard deviation equal to $SD = 0.005$. After transformation of this dependence to a form of Eq. (2.16) the values of g and g_0 were finally obtained as

$$g = 7.15 \pm 0.16, \quad g_0 = 0.23 \pm 0.01. \tag{2.20}$$

The critical isotherm is usually described employing Eq. (1.10). However, as in this case every time only small parts of it were examined, it was possible to use only the main asymptotic term from Eq. (1.10). In Fig. 2.29 the correlation between critical amplitude (D_0) and index (δ) is shown in a log–log scale for the critical isotherms of some pure substances. It was plotted in the same way as previously. The straight line shown on the graph (Fig. 2.29) is plotted using "experimental" values of 23 pairs D_0 and δ, and satisfies the linear dependence with a correlation coefficient $R = 0.997$ with a standard deviation equal to $SD = 0.04$. After transformation of this dependence to a form of Eq. (2.17), the following values of d and d_0 were finally obtained:

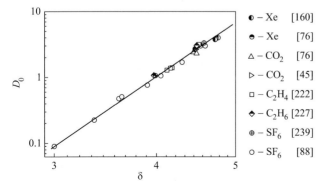

Figure 2.29 Correlation between the critical amplitude (D_0) and index (δ) for the critical isotherm [237].

$$d = 8.48 \pm 0.15, \quad d_0 = (8.2 \pm 1.5) \times 10^{-6}. \tag{2.21}$$

Now, after having established the values of all the necessary constants it becomes possible to calculate the universal combination R_χ^+.

However, the values of three critical indices, connected by the universal relation (2.4), should be fixed before. It is possible to do this logically consistently in two ways. Firstly, it could be a theoretical set which satisfies the three-dimensional Ising model. In this case, we have the well-known set of parameters: $\beta = 0.326$, $\gamma = 1.239$ and $\delta = 4.80$ (see, e.g., [26, 52]). Secondly, it could be possible to use values obtained independently in the same experiment, with the same highly pure sample. For this purpose, naturally, our choice falls again on our $(p\rho T)$-experiment on SF_6, which was discussed in Chapter 1, as it fully satisfies all requirements demanded of such an experiment.

The substitution of a set of critical indices into Eq. (2.18) corresponding to the Ising model together with the constants found (Eqs. (2.19)–(2.21)) gives $R_\chi^+ = 0.96$, which differs significantly from the theoretical one. This fact, once more, demonstrates that critical index values, which satisfy this model, can be seen only as a zeroth-order approximation for the situation typical for real liquids. With the second method, taking into account Eq. (2.12) and Eqs. (2.19)–(2.21), we get $R_\chi^+ = 1.44$, which is close both to the theoretical RG-value ($R_\chi^+ = 1.6$) and to $R = 1.48 \pm 0.05$, which was obtained by the direct substitution of the corresponding amplitudes and conjugated critical index values from Eq. (2.12) into the relation for R_χ^+, Eq. (2.13).

Thus, the results obtained in the course of this analysis confirm the suggestion about the existence of correlation dependences like Eqs. (2.15)–(2.17) between critical amplitudes and indices, which are involved in the theoretical description of the critical behavior (Eq. (1.2)) for various physical properties of pure classical liquids. The universality of critical indices makes it possible, on this basis, to also consider critical amplitudes as "quasiuniversal" within the limits of a certain class of matter. In order to find out whether such features exist for other groups of matter, included in the $(d = 3, n = 1)$-class of universality, further analyses are required.

It should be emphasized that the presence of such correlation dependences has also a significant heuristic value and makes the estimation system of experimental data quality more complete. It turned out that the agreement of critical indices, even if independently determined, with the theoretical equation (2.4) is necessary, but totally insufficient. It is also required that the amplitudes satisfy the empirical relations (2.15)–(2.17). In the rather frequent cases when only separate sections of the thermodynamic surface are studied and, therefore Eq. (2.4) cannot be used, to satisfy these correlation dependences is, in essence, the only criterion of the obtained experimental data self-consistency. If the wish arises, during the experiment to enter deeper into the critical region, it is important to remember in this case that when carrying out such a check it is necessary to be particularly careful not to attract into the analysis experimental data belonging to the region after the second crossover [2, 164, 165, 226, 237], as this will lead to an incorrect determination of critical indices and, especially, amplitudes.

3

Thermodynamics of the Metastable State

3.1
The "Pseudospinodal" Hypothesis

The topic of this chapter, although not directly related to the main theme of the whole book, is devoted to a question which is by no means secondary. What we are talking about here is whether the critical point is unique as a point, where the singularity of the fullest set of thermophysical properties appears simultaneously, or is it an ordinary representative of a whole family of such points, forming a special line, the pseudospinodal. Of course, we should not also ignore the more general question posed in the title of the chapter.

Despite the large number of books and articles which have already been devoted to the domain of metastability (e.g., [241–243]) it still remains a certain *terra incognita* in liquid state physics. This state of affairs, unquestionably, leads to the occasional appearance of various hypotheses, some of which are clearly quite unrealistic. Within the limits of one small chapter our aim is not even to try to comprehend all details of this subject very deeply. Our task is much more modest. That is to discuss a set of principle considerations concerning the once so popular so-called pseudospinodal hypothesis, which suggests the existence of a certain line, the "pseudospinodal," whose all points possess the properties of the critical one [244]. These considerations also touch on the question of the very existence of such a universal "pseudospinodal" curve [141, 245].

3.1.1
The History of the Occurrence of the "Pseudospinodal Hypthesis"

In 1969, George B. Benedek, one of the pioneers of correlation spectroscopy [246], published a review [247] on the application of this method in physics, chemistry, biology, and engineering (about essential features of correlation spectroscopy see, in particular, Chapters 4–6 and the Appendix). At that time correlation spectroscopy had greater future potential rather than actual successes, which made the review particularly timely. In it, Benedek presented an analysis of detailed experimental research on the behavior of the thermal conductivity of SF_6 along different lines close to the critical point. The analysis was carried out

Critical Behavior of Nonideal Systems. Dmitry Yu. Ivanov
Copyright © 2008 WILEY-VCH Verlag GmbH & Co. KGaA, Weinheim
ISBN: 978-3-527-40658-6

by Benedek's colleague J. B. Lastovka. We mention this research, first because its results showed an unusually weak singularity of SF_6 thermal conductivity. It was just after this research that the necessity of taking into account background, nonsingular, parts of kinetic coefficients became quite obvious (for more details, see Chapter 7).

Among mentioned and other things, a very important observation was made during this analysis. It was found that the character of the C_p-singularity along both the critical and noncritical isochores appeared to be the same. This feature helped to find the spinodal position by extrapolation.

Almost simultaneously to this review [247] an article was published (Chu et al. [248]) in which the authors, referring to Benedek, introduced, for the first time, the idea of the "pseudospinodal." In this work it was clearly emphasized that the use of the term "pseudospinodal" instead of "spinodal" was caused by the fact that "in contrast to the pseudospinodal, a line defined by extrapolating from stable domains, the position of the true spinodal (if it exists) could only be determined by special experiments in the metastable region." We have to mention that the authors of the article [248] could not have read Benedek's work itself as it had not yet seen the light of day. They could only have known about its results, as often happens, either from Benedek himself or from a preprint sent to them. Therefore, the reference to his work was entered very approximately (1968 instead of 1969 and the title of the French collection where the work was published was also inaccurately cited). It is interesting that even after the paper [247] had been published, articles [249–252] appeared in which the author of the "pseudospinodal" was still considered to be Benedek himself. Note that Benedek was not talking at all about the "pseudospinodal" or the singularities of other thermodynamic quantities (except for C_p) in his work [247].

3.1.2
The Universal "Pseudospinodal"

All that was said above about the pseudospinodal, although indicating a certain carelessness concerning the authors references to quotes from primary sources, is related exclusively to "priority-terminological" questions without touching on the essence of the problem. The authors of the paper [244], however, went even further. They "attributed" to Benedek not only the "pseudospinodal" but also the universal "pseudospinodal," on which "all thermodynamic and transport properties diverge just as on the critical point." So according to Benedek, or in fact according to Osman and Sorensen [244], the pseudospinodal is defined by the equation $X = X_0[T^* - T^*_{sp}(\rho)]^{-x}$, where X is the investigated thermodynamic or transport property, and X_0, x, T^*, and $T^*_{sp}(\rho)$ are the critical amplitude and index, temperature, and "pseudospinodal" temperature (made dimensionless with the help of T_c), respectively. The set of $T^*_{sp}(\rho)$ points forms the universal "pseudospinodal" curve. In their work, Osman and Sorensen [244] claimed that they had succeeded, by using a huge amount of various experimental data for different substances, from CO_2 to ^4He, in convincingly demonstrating that such a line really exists. Below, we will show that the existence of such a line would contradict the basic principles of thermodynamics.

3.2
The van der Waals Spinodal

Supersaturated vapor and superheated liquid are particular examples of the metastable state of matter. The region of metastable states in the phase diagram is bounded by two lines, the binodal and the spinodal curves. The binodal (coexistence curve, boundary curve) separates metastable and absolutely stable homogeneous states. The spinodal is the boundary between metastable and unstable states, the latter, as a rule, quickly lose their homogeneity as they are unstable with respect to small amplitude but long wavelength perturbations in density and/or composition [241–243, 253].

The van der Waals equation, despite its simplicity, is surprisingly correct in a qualitative sense and reproduces all the basic features of single- and two-phase states of matter, the metastable region included. Unfortunately, as is well known, it is not realistic to obtain from this really remarkable equation a quantitative description close to the critical point. However, as shown in the two preceding chapters, it seems that in the close vicinity of the critical point the mean-field theory and its special case, the van der Waals equation, can be applied once again.[1]

It is not difficult to show that in the van der Waals theory the isothermal compressibility, K_T, the isobaric heat capacity, C_p, and the isobaric expansion coefficient, α_p, become infinite on the spinodal, while the isochoric heat capacity, C_V, is finite everywhere. Actually, as the spinodal goes through extrema of isotherms in (p, V)-coordinates, the necessary condition determining the location of the spinodal can be expressed as

$$\left(\frac{\partial p}{\partial V}\right)_T^{(sp)} = 0. \tag{3.1}$$

This condition together with the van der Waals equation for one mole of a substance,

$$p = \frac{RT}{V - b} - \frac{a}{V^2}, \tag{3.2}$$

allows one to determine the spinodal. In the dimensionless form, we can rewrite the van der Waals equation as

$$p^* = \frac{8T^*}{3V^* - 1} - \frac{3}{(V^*)^2}, \quad p^* = \frac{p}{p_c}, \quad T^* = \frac{T}{T_c}, \quad V^* = \frac{V}{V_c}, \tag{3.3}$$

where p^*, T^*, and V^* are nondimensional pressure, temperature, and molar volume, respectively. With these notations, the spinodal equation can be presented in different coordinates as given below:

---

[1] Here it seems quite appropriate to paraphrase Einstein ("Gott ist raffiniert, aber er ist nicht bösartig") as "The Lord is not malicious, but subtle," remembering his famous phrase given in the original version earlier and also below.

$$T_{sp} = \frac{2a(V-b)^2}{RV^3}, \quad T_{sp}^* = \frac{(3V^*-1)^2}{4(V^*)^3}, \tag{3.4}$$

$$p_{sp} = \frac{a(V-2b)}{V^3}, \quad p_{sp}^* = \frac{3V^*-2}{(V^*)^3}. \tag{3.5}$$

These equations allow one to construct the van der Waals spinodal. The spinodal shape corresponds to its name: *spina (lat.)*: thorn (see, e.g., [243], pp. 21 and 23). The term was introduced by van der Waals.

Let us examine, now, the behavior of isothermal compressibility on the spinodal. According to the definition of isothermal compressibility and Eq. (3.1) regardless of the chosen equation of state, we immediately get

$$K_T \equiv -\frac{1}{V}\left(\frac{\partial V}{\partial p}\right)_T^{(sp)} = \infty. \tag{3.6}$$

Following the definition of the thermal expansion coefficient, let us differentiate Eq. (3.2) with respect to temperature at constant pressure. Then taking into account Eq. (3.4) we get

$$\alpha_p = \frac{1}{V}\left(\frac{\partial V}{\partial T}\right)_p^{(sp)} = \frac{1}{V}\frac{R(V-b)}{RT/(V-b)^2 - 2aV^{-3}} = \infty. \tag{3.7}$$

3.2.1
First-Order Stability Conditions

As far as the spinodal is the stability boundary for continuous phase changes [241, 243] it is necessary, for further examination, to look at the conditions which determine the thermodynamic stability of the system. We assume that the necessary equilibrium conditions, such as temperature and pressure equality in different parts of the system, are satisfied. So for thermodynamic stability of the system it is necessary that inequality (3.8), following from maximum entropy [254] or minimum internal energy [255] in the equilibrium state, is satisfied in its strict sense ([256], p. 110)

$$\delta U + p\delta V - T\delta S > 0. \tag{3.8}$$

This requirement, in turn, will be satisfied if the second-order variation of the internal energy is essentially positive [254–258]

$$\delta^2 U = \frac{1}{2}\left(\frac{\partial^2 U}{\partial V^2}\delta V^2 + 2\frac{\partial^2 U}{\partial V\delta S}\delta V\delta S + \frac{\partial^2 U}{\partial S^2}\delta S^2\right) > 0. \tag{3.9}$$

Let us clarify the necessary and sufficient conditions under which the quadratic form given by Eq. (3.9) will always be positive. Therefore, let us examine the identity which is analogous to Eq. (3.9), i.e.,

$$U(x,y) \equiv ax^2 + 2cxy + by^2 \equiv \frac{1}{a}\left[(ax+cy)^2 + y^2(ab-c^2)\right], \tag{3.10}$$

from where we immediately get the desired result for the stability determinant

$$D \equiv \left(\frac{\partial^2 U}{\partial V^2}\right)\left(\frac{\partial^2 U}{\partial S^2}\right) - \left(\frac{\partial^2 U}{\partial V \partial S}\right)^2 > 0,$$

$$\frac{\partial^2 U}{\partial V^2} > 0 \qquad \frac{\partial^2 U}{\partial S^2} > 0.$$

$$(3.11)$$

Applying the thermodynamic identity $TdS = dU + pdV$, taking into account Eq. (3.11), we obtain

$$U_{SS} \equiv \frac{\partial^2 U}{\partial S^2} = \left(\frac{\partial T}{\partial S}\right)_V = \frac{T}{C_V} > 0,$$

$$U_{VV} \equiv \frac{\partial^2 U}{\partial V^2} = -\left(\frac{\partial p}{\partial V}\right)_S = \frac{1}{V K_S} > 0,$$

$$U_{VS} = U_{SV} = -\left(\frac{\partial p}{\partial S}\right)_V = \left(\frac{\partial T}{\partial V}\right)_S.$$

$$(3.12)$$

Using standard methods of transformation of thermodynamic expressions, in particular, the method of Jacobi determinants (see, e.g., [258]), it is not difficult to express the stability determinant D using the quantities that we are interested in. We get

$$D = U_{SS} U_{VV} - U_{SV} U_{VS} = \frac{\partial(T, -p)}{\partial(S, V)} = \frac{\partial(T, -p)}{\partial(T, V)}\frac{\partial(T, V)}{\partial(S, V)} = \frac{T}{C_V V K_T},$$

$$D = U_{SS} U_{VV} - U_{SV} U_{VS} = \frac{\partial(T, -p)}{\partial(S, V)} = \frac{\partial(T, -p)}{\partial(S, p)}\frac{\partial(S, p)}{\partial(S, V)} = \frac{T}{C_p V K_S}.$$

$$(3.13)$$

Then, for stable states, we get

$$\frac{T}{C_V} > 0, \qquad \frac{1}{V K_T} > 0,$$

$$\frac{T}{C_p} > 0, \qquad \frac{1}{V K_S} > 0,$$

$$(3.14)$$

where K_S, the adiabatic compressibility, is determined as

$$K_S \equiv -\frac{1}{V}\left(\frac{\partial V}{\partial p}\right)_S.$$

$$(3.15)$$

3.2.2
Higher Order Stability Conditions

On the spinodal, where $D = 0$ holds [241–243], some of the first-order stability conditions (3.14) are violated. The inequality sign changes to an equality sign. As a result, on the spinodal C_p and K_T become, at the same time, infinitely large. According to Gibbs, $D = 0$ holds also at the critical point, where also $dD = 0$ ([256], p. 135). The stability matrix determinant, which can be determined from the conditions of minimum of the Gibbs thermodynamic potential (see, e.g., [259], pp. 265–266),

$$\begin{vmatrix} \Delta T & \Delta p \\ \Delta V & \Delta S \end{vmatrix} > 0, \tag{3.16}$$

gives, in this case, the possibility of arriving at a set of stability conditions of an already higher order

$$\left(\frac{\partial^2 p}{\partial V^2} \right)_T = 0, \quad \left(\frac{\partial^3 p}{\partial V^3} \right)_T < 0, \quad T = \text{const.}$$

$$\left(\frac{\partial^2 T}{\partial S^2} \right)_p = 0, \quad \left(\frac{\partial^3 T}{\partial S^3} \right)_p > 0, \quad p = \text{const.} \tag{3.17}$$

Therefore, in general, the stability conditions are given by the requirements that the first nonzero derivative with respect to pressure (temperature) should be, first of all, of odd order and secondly negative (positive).

3.2.3
Approaching the Instability Points

Inequalities (3.14) show that for stable states all four inverse to C_p, K_T, C_V, and K_S quantities are positive. As for the thermal expansion coefficient (3.7), it can be presented as

$$\alpha_p = \frac{1}{V} \left(\frac{\partial V}{\partial T} \right)_p = K_T \left(\frac{\partial p}{\partial T} \right)_V . \tag{3.18}$$

Since $(\partial p / \partial T)_V$ is positive everywhere and, as experiment shows (see, e.g., [91,126] and Fig. 1.21), nonsingular α_p has the same sign, the same zeros, and "infinities" as K_T. Therefore, by virtue of the well-known thermodynamic relations (see, e.g., [26], pp. 62–65)

$$C_p = C_V + \frac{TV}{K_T} \alpha_p^2, \tag{3.19}$$

$$K_T = K_S + \frac{TV}{C_p} \alpha_p^2, \tag{3.20}$$

$$\frac{K_T}{K_S} = \frac{C_p}{C_V}, \tag{3.21}$$

for the stable phases the following inequalities are satisfied:

$$C_p > C_V > 0, \quad \frac{T}{C_V} > \frac{T}{C_p} > 0, \quad K_T > K_S > 0, \quad \frac{1}{VK_S} > \frac{1}{VK_T} > 0. \tag{3.22}$$

From these inequalities it follows that in the absence of infinite values of (T/C_V), stability is probably violated first of all when the quantities (T/C_p) and $(1/(VK_T))$ simultaneously go through zero. However, if C_p and K_T go through zero, then C_V and K_S, under these conditions, approach zero faster ([259], p. 232).

3.2.4
The Instability Area

Let us assume that the state of the system is represented by points on the critical isochore $(V = 3b)$ in the two-phase area $(T < 8a/27Rb)$. From van der Waals' equation, we get

$$\left(\frac{\partial p}{\partial T}\right)_V = \frac{R}{V - b} > 0 \quad \text{for all} \quad V > b,$$

$$\left(\frac{\partial p}{\partial V}\right)_T = -\frac{RT}{(V - b)^2} + \frac{2a}{V^3}.$$

(3.23)

In fact, this means that the "T"-derivative, as was already mentioned, is nearly always positive. Then, for the selected state variables by Eq. (3.23), and taking into account Eqs. (3.6) and (3.7), for constant density we get

$$\left(\frac{\partial p}{\partial V}\right)_T > 0, \quad \frac{1}{VK_T} < 0, \quad \alpha_p < 0.$$

(3.24)

As is known, in the mean-field theory, whose special case is the van der Waals equation, the isochoric heat capacity is positive and finite everywhere outside the critical point, where it undergoes a jump (see Eq. (3.24)). Then, according to Eq. (3.19), $C_p - C_V < 0$ holds. This inequality means that C_p can, in principle, have any sign and that the sign of K_S is opposite to the sign of C_p (see Eq. (3.21)). Along any isotherm in a two-phase area the derivative $(\partial p/\partial V)_T$ goes through zero twice, thus fixing the spinodal, which together with the condition $C_V > 0$ determines the limits of stability. On the spinodal the stability conditions $(T/C_p) > 0$ and $(1/(VK_T)) > 0$, as we have already noted, are violated and both these quantities $(C_p$ and $(1/(VK_T)))$ go through zero. So, if somewhere in the instability region, the heat capacity C_p is positive, then that part of the area where $C_p > 0$ holds is limited by zero values of C_p [259]. In particular, C_p has zero values on the so-called adiabatic spinodal (where $D < 0$ [243]). This line is formed by the zeros of the inverse adiabatic compressibility K_S^{-1}, and at the same time the "common" spinodal coincides with the zeros of K_S itself.

The fact that phase stability is usually violated when both (T/C_p) and $(1/(VK_T))$ go through zero simultaneously probably explains the practice of using only two stability conditions: $(T/C_p) > 0$ and $(1/(VK_T)) > 0$, instead of the whole set (3.14) while searching for critical points or the spinodal equation (see, e.g., [243]). So, we can see that, at least within the framework of the van der Waals equation, if K_T, C_p, and α_p become infinite on a common line, the spinodal (these quantities also go through zero together), then the adiabatic compressibility, K_S, in contrast, has a zero value on the spinodal and becomes infinite on a completely different curve, on the adiabatic spinodal.

Note, the analysis carried out here essentially is a generalization of a number of known thermodynamic problems, which use the van der Waals equation as an equation of state, applied to the problem under examination (see, e.g., [259]). In fact, the choice of this particular equation of state, as can be seen, in no way limits the generality of the conclusions arrived at as a result of the analysis. This analysis, together with the reliably established fact that C_V is infinite at the critical point, convincingly shows, in our opinion, that a universal line, whatever it is called, on which the same quantities could diverge, that are singular in a critical point, does not exist.

To conclude this section it should be noted that the analysis shows that neither the van der Waals spinodal nor any other line can perform the role of a universal curve "in the spirit of Osman and Sorensen" [244]. The very process of this analysis did not leave any doubt that, in all probability, the use of a specific equation should not stop us from making more general conclusions about features which appear in the metastable area. Nevertheless, as in the van der Waals theory C_V does not have infinite values anywhere, including the critical point [25, 27], then in order to get greater generality and strictness it was necessary to involve the now well-known experimental and theoretical result which is beyond the van der Waals theory as well, namely, the singularity of C_V at the critical point. In fact, the conclusion that the singularity lines of K_T, C_p, and α_p, on the one hand, and of K_S, on the other, are different lines does not bear any relation to what was said above.

This conclusion remains valid. Only the isochoric heat capacity C_V does not fit the general picture, as after such a deviation from "classical" behavior as the C_V singularity at the critical point was discovered, the possibility, maybe only hypothetical, appeared for it to remain infinite outside this point on a certain line. Within the limits of the discussed suggestions [244] it was considered that this line would be similar to the spinodal. At a first glance such a hypothesis seems attractive: it seems to take us in a qualitative sense back to the van der Waals theory, where the critical point is an ordinary representative of the spinodal curve. (Here the term "ordinary" is appropriate, obviously, only from the point of view of the joint divergence of thermodynamic quantities. The fact that the critical point is the only point on the whole thermodynamic surface belonging at the same time to the binodal, spinodal, and critical isotherm makes it totally exceptional, even by formal characteristics.) It is not possible, however, to make such a hypothesis agree with the principles of thermodynamics, not only when using a specific type of state equation, in particular the van der Waals equation, but also, as will be shown in the next section, in the most general case [245].

3.3
Thermodynamic Analysis of the "Pseudospinodal" Hypothesis

3.3.1
Physics and Geometry

One of the distinctive features widely used in his thermodynamical works by J. W. Gibbs, one of the founders of modern thermodynamics, who extended it to multiphase, multicomponent, and heterogeneous systems, was geometry. In the cycle of three thermodynamic papers (even the titles of some of the works of this cycle are typical: "Graphical methods in the thermodynamics of fluids," "A method of geometrical representation of the thermodynamic properties of substances by means of surfaces" [256]), Gibbs showed how, with the help of geometrical methods, we can analyze the behavior of different systems and study the limits of their thermodynamic stability and critical phenomena therein. This connection may seem surprising only at a first glance. Many, if not the majority, of properties of the world which surround us can be interpreted geometrically: physics geometry of space–time in special and general relativity theories, physics geometry of order – disorder phenomena, phase transition physics – scale invariance – ε-expansion – fractal geometry, Brownian trajectory – polymer chains – multiple light (neutron) scattering (the connection between "Brownian trajectory – polymer chain – multiple light scattering" will be discussed in Chapters 5 and 6); geography – cosmology – fractal geometry.

This list is far from being a complete one and it could be easily extended. We just wanted to underline here the universality and fundamentality of geometric concepts in nature (see, e.g., [260, 261]). The only difficulty consists in the plurality of geometries themselves and in the choice of an adequate geometric conception in the framework of any of them.

3.3.2
Mathematical Foundation

The "universal pseudospinodal" hypothesis will be further discussed in connection with the work [245] using the most general mathematical concepts, allowing us to give a fairly simple geometric interpretation. The first and second principles of thermodynamics establish a connection between thermal and caloric properties of matter and set the direction for the development of natural processes. As a consequence of both thermodynamic principles there exist thermodynamic potentials, each of which has its own natural set of variables. For variables (T, V), such a potential is the Helmoltz free energy, F. Below, we shall show that the suggestion about the simultaneous divergence to infinity of C_p, K_T, and C_V results in so many contradictory requirements concerning this surface, $F(T, V)$, that it is impossible to satisfy them.

To prove our assertions we first need to summarize some mathematical facts. First of all, let us discuss the behavior of the first derivatives of a continuous function, f. Let $f(x, y)$ be continuously differentiable within the area G, while the boundary of this area, L, is the line of singularity of one of the first derivatives: when $(x, y) \in L$, then $(\partial f / \partial x)$ becomes infinite; however, $f(x, y)$ is finite and continuous there. Let us suppose that at a certain point $(x_0, y_0) \in L$ there exists the derivative $(dy/dx)_{sin}$, so that

$$\left(\frac{dy}{dx}\right)_{sin} \neq 0, \quad \left(\frac{dx}{dy}\right)_{sin} \neq 0. \tag{3.25}$$

The index "sin" means that differentiation takes place along the singularity line L. Let us choose the point (x_1, y_1) on L in such a way that $(x_0, y_0) \in G$. Further we use the identity

$$\frac{f(x_1, y_1) - f(x_0, y_1)}{x_1 - x_0} = - \frac{y_1 - y_0}{x_1 - x_0} \frac{f(x_0, y_1) - f(x_0, y_0)}{y_1 - y_0}$$

$$+ \frac{f(x_1, y_1) - f(x_0, y_0)}{x_1 - x_0}. \tag{3.26}$$

Now we go over to the limit $x_1 \to x_0$. The left-hand side of Eq. (3.26) becomes infinite and, if demanded that for $(x_1, y_1) \in L$ the relation

$$\lim_{\varepsilon \to 0} \sup_{|x_1 - x_0| < \varepsilon} \left| \frac{f(x_1, y_1) - f(x_0, y_0)}{x_1 - x_0} \right| \neq \infty \tag{3.27}$$

holds, then the second term on the right-hand side, thanks to condition (3.27), remains finite with respect to its absolute value. Consequently, near the singularity line, we have

$$\frac{\partial f}{\partial x} \sim - \left(\frac{dy}{dx} \right)_{\sin} \frac{\partial f}{\partial y}. \tag{3.28}$$

The obtained asymptotic correlation is easy to generalize when, on L, the first-order derivatives from $F(x, y)$ are continuous, but the second derivative $(\partial^2 F / \partial x^2)$ becomes infinite. Assuming that the conditions given by Eqs. (3.25) and (3.27) are satisfied for $f = (\partial F / \partial x)$ and $f = (\partial F / \partial y)$, then close to the singularity line we get

$$\frac{\partial^2 F}{\partial x^2} \sim - \left(\frac{dy}{dx} \right)_{\sin} \frac{\partial^2 F}{\partial x \partial y} \sim \left(\frac{dy}{dx} \right)_{\sin}^2 \frac{\partial^2 F}{\partial y^2}. \tag{3.29}$$

All further conclusions are based on these asymptotic relations.

Relations (3.29) allow us to confirm that the second-order derivatives

$$\frac{\partial^2 F}{\partial x^2}, \quad \frac{\partial^2 F}{\partial x \partial y}, \quad \frac{\partial^2 F}{\partial y^2} \tag{3.30}$$

become infinite on a common line. In other words, second-order derivatives should diverge together. Note that this statement is inapplicable to those points where conditions (3.25) are violated, i.e., where the tangent to the singularity line is parallel to one of the coordinate axes (at such points one of the second derivatives can be infinite, while two others are finite). We would also like to mention that the limitations imposed by Eq. (3.27) are practically superfluous: the functions $(\partial F / \partial x)$ and $(\partial F / \partial y)$ are continuous on the singularity line, but the continuity of the function f on a certain line means that the points, where condition (3.27) is violated, form on this line a null set (more often, these points do not exist at all).

3.3.3
Thermodynamic Consequences

Equations (3.29) can be applied to any appropriate thermodynamic potential. Let us begin with Helmholtz's free energy $F(T, V)$, for which

$$dF = -SdT - pdV \qquad (3.31)$$

holds. Then, close to the singularity line of second-order derivatives we obtain

$$-\frac{C_V}{T} \sim \left(\frac{dV}{dT}\right)_{\text{sin}} \left(\frac{\partial p}{\partial T}\right)_V \sim \left(\frac{dV}{dT}\right)_{\text{sin}}^2 \frac{1}{VK_T}. \qquad (3.32)$$

So, if on one line $C_V = \infty$ is fulfilled, then on the same line $K_T = 0$ holds. Consequently, the universal "pseudospinodal" cannot exist.

It should be stressed that at the critical point one of the conditions (3.25) is violated; therefore, this is the only point where K_T and C_V can diverge together. From Eq. (3.32) it follows that the singularity line cannot be a spinodal, too. In fact, the spinodal is the boundary between metastable and unstable states. For metastable states, the stability conditions (3.11) and (3.14) should be satisfied, i.e., $D > 0$ and $C_V > 0$ should hold. However, from Eq. (3.32) it follows that K_T and C_V have opposite signs; therefore, close to the C_V singularity line the determinant D is less than zero. As a consequence, the line on which $C_V = \infty$ is surrounded by unstable states, and metastable states are not close to it.

The spinodal is a singularity line of second-order derivatives from Gibbs' thermodynamic potential $G(T, p)$, for which

$$dG = -SdT + Vdp \qquad (3.33)$$

holds. If in Eq. (3.29) we change F to $-G$, then these relations turn into

$$\frac{C_p}{T} \sim \left(\frac{dp}{dT}\right)_{\text{sin}} V\alpha_p \sim \left(\frac{dp}{dT}\right)_{\text{sin}}^2 VK_T. \qquad (3.34)$$

So, at the spinodal K_T, C_p, and α_p diverge together. As for C_V, comparison of Eqs. (3.32) and (3.34) leads to the conclusion that C_V and C_p cannot become infinite on the same line. And this means that C_V is always finite on the spinodal.

Further, for enthalpy $H(S, p)$, we have

$$dH = TdS + Vdp. \qquad (3.35)$$

Therefore, close to the singularity line of its second-order derivatives, we get

$$\frac{T}{C_p} \sim -\left(\frac{dp}{dS}\right)_{\text{sin}} \left(\frac{\partial T}{\partial p}\right)_S \sim -\left(\frac{dp}{dS}\right)_{\text{sin}}^2 VK_S. \qquad (3.36)$$

On this line, adiabatic compressibility diverges, which is the reason why it is called the adiabatic spinodal [241, 243]. The comparison of Eqs. (3.34) and (3.36) shows that the simultaneous divergence of K_T and K_S is not possible on the line. From Eq. (3.36) it follows that on the adiabatic spinodal $C_p = 0$ and C_p and K_S have opposite signs, i.e., $C_p K_S < 0$. Thus, we can see from Eq. (3.13) that close to the adiabatic spinodal the determinant $D < 0$, i.e., this line is surrounded by unstable states. All these conclusions were also established within the framework of the previous, less general, analysis (see Section 3.1).

Finally, if there exists a singularity line of second-order derivatives of the internal energy, $U(S, V)$, then close to this line the relation

$$\frac{T}{C_V} \sim -\left(\frac{dV}{dS}\right)_{\text{sin}}\left(\frac{\partial T}{\partial V}\right)_S \sim -\left(\frac{dV}{dS}\right)_{\text{sin}}^2 \frac{1}{VK_S} \tag{3.37}$$

is fulfilled. From this relation it follows that C_V can become zero only together with K_S. However, the suggestion that such a line exists, apparently, cannot be in agreement with the usual understanding about the behavior of matter in the metastable state.

The main result of the performed analysis [141, 245] is that the restrictions placed by thermodynamic principles on the simultaneous singular behavior of thermodynamic coefficients close to a particular line appear significantly stricter than analogous restrictions close to an isolated particular point. We can distinguish a group of thermodynamic coefficients, which diverge together. Such a group is formed by second-order derivatives of one and the same thermodynamic potential. At the same time, there exist pairs of thermodynamic coefficients (K_T and C_V, C_V and C_p, C_p and K_S, K_S, and K_T), for which a simultaneous transformation both to zero and to infinity is forbidden. These prohibitions can be violated at individual points; however, such points can never form a line. In particular, such unique properties are inherent to the critical point where K_T, C_p, and C_V diverge together, although such a behavior is impossible on a line.

3.4
Experimental Test of the "Pseudospinodal" Hypothesis

Before the publication of the paper [245], where a complete thermodynamic analysis of the "pseudospinodal" hypothesis was first carried out, this hypothesis was analyzed repeatedly with respect to the question whether it agrees with the scale invariance hypothesis or not (see, e.g., [262, 263]). As we mentioned in the first chapter, the idea of scale invariance, or scaling, which was formulated in the mid-1960s, is one of the most productive ideas in the physics of phase transitions and critical phenomena [58–62]. Scaling is based on the inviolable physical fact that close to the critical point the correlation radius significantly exceeds molecular size, and therefore this parameter can be taken as the only distinctive scale in the system. Scale invariance implies that thermodynamic potentials are homogeneous functions of their variables (see, e.g., [26]); this in turn leads directly to the establishment of relations between critical indices (see, e.g., [52]), only two of which are independent, and the space dimension. The concrete values of critical indices also depend on the type of symmetry of the system under investigation. This is what determines the existence of different universality classes.

It would not really be a great exaggeration to state that the basic successes in describing critical phenomena are, in one way or another, related to the idea of scale invariance. Although, as Stanley writes in his book ([26], p. 54), "at best it has not been proven." Of course, this statement was made still in the "prerenormalization group" era. Applying, as it is done now, renormalization group ideas, the block Hamiltonian, as it does in scaling, has a central position (see, e.g., [52]). To calculate the critical indices we must turn to space of fractional dimension, $d = 4 - \varepsilon$, within the framework of the so-called ε-expansion (see, e.g., [51]). Fractional dimension is in this case similar to fractal dimension, the central quantity in fractal geometry, which was mentioned above. The successes of the theory, achieved in this direction over the last few years, are fully confirmed by accurate theoretical predictions, obtained by different methods, which closely agree between themselves and with experiments [40, 50].

Therefore, we can confidently assume that scale invariance is now not only just an established fact, but also a working instrument of critical phenomena theory. This is the reason why attempts to verify the "pseudospinodal" hypothesis with scaling concepts are very reasonable. Naturally, all the attempts gave negative results. Naturally, because, as the analysis (see Section 3.3) shows, the prohibition of the existence of a universal "pseudospinodal" comes directly from thermodynamic principles. Consequently, this hypothesis, by necessity, should come into conflict with any consistent thermodynamic theory.

Since in [244], where this hypothesis was put forward, it was asserted that the experimental data used for illustration did not contradict it, experiments were carried out in different laboratories and on different substances specifically to verify it. For example, Skripov and colleagues [264] studied the behavior of the isochoric heat capacity of liquid xenon. The tests were carried out close to the liquid–vapor critical point, where one might expect to find prespinodal peculiarities even for shallow penetration into the metastable area. Absolute errors in the measurements of heat capacity were of the order ~5%, and relative ones did not exceed 1%. In going through the line of phase equilibrium with penetration into the metastable area along two isochores, differing by 23 and 32% from the critical, no anomalies in the temperature dependence of C_V or prespinodal peculiarities were discovered.

Matizen and colleagues also carried out special research on the isochoric heat capacity of liquid ^{4}He [265, 266] using both their own experimental data and the results of Moldover's group. Previously, the whole data array was checked for self-consistency. Without going into the details of these painstakingly and delicately carried out works, we arrive at the final conclusion, taking into account its importance: "^{4}He data processing does not allow us to draw a conclusion on the possibility of describing C_V on noncritical isochors by scaling type formulae, where the reduced temperature is counted off from the spinodal curve" [265].

So, the attempt to introduce into scientific use the so-called pseudospinodal hypothesis [244] to describe the metastable area, was unsuccessful. This idea could not be made to agree either with the principles of thermodynamics, or with the theory of scale invariance. Specially carried out experiments were also unable to confirm this. As a result, it can be stated that the *existence of a universal line on which all the same thermodynamic and transport properties, which diverge at the critical point, would simultaneously diverge is prohibited by thermodynamics. The critical point is the only point of the thermodynamic surface where this is possible.*

Therefore, the results of this chapter, apart from anything else, are extremely important ideologically. They establish the fact that what is implied by the term "critical point" is not related to one of the many possible points on the thermodynamic surface, but to the unique point which simultaneously lies on the binodal, the spinodal, and the critical isotherm.[2]

2) It is once more appropriate to remember Einstein, but this time not paraphrased "Subtle is the Lord, but malicious He is not."

Part II
The Dynamics of Critical Phenomena

4

Foundations of Critical Dynamics

4.1
Introduction

It is an anomalous growth in order-parameter fluctuations (density, concentration) close to the critical point that lies at the heart of critical dynamics. According to Onsager's hypothesis [267], the process of fluctuation decay is described by the same hydrodynamic equations as for corresponding macroscopic impacts. Therefore, the dynamics of critical phenomena is naturally determined by the time evolution of these fluctuations. The anomalous behavior of kinetic coefficients close to the critical point is caused by the slowing down of large-scale order-parameter fluctuations as they get closer to the phase transition point. A simple picture of the phenomena accompanying this "critical slowing down" is a "drop," whose radius grows as the correlation length, and which performs a Brownian motion. This "droplet" is a large-scale entropy fluctuation. As the critical point is approached, because of the growth of the "droplet's" radius, its movement, and therefore, the rate of heat flow decrease. Thermal diffusivity, which determines the rate of decay of these fluctuations, decreases toward zero at the critical point, despite the fact that thermal conductivity shows anomalous growth in this region (see, e.g., [40]). The classical kinetic theory of dense gases and liquids does not suggest any anomalies for kinetic coefficients close to the critical point [268, 269]. This is natural as such a theory, which examines the collisions between different molecules (mostly paired), does not take into account the cooperative character of their behavior in the neighborhood of the phase transitions points.

Van der Waals was the first to understand (see, e.g., [14, 15]) that the critical behavior of pure liquids is, in fact, a direct consequence of a "strong" singularity of isothermal compressibility (in a more general sense, of the susceptibility). The van der Waals theory [15], like all other classical critical phenomena theories, takes into account the cooperative character of the transition, but neglects fluctuations and assumes the possibility of a series expansion of a thermodynamic potential with respect to the order parameter. Therefore, all these theories, in one way or another, lead to a mean-field theory. However, the flaw in the theories of this type consists not mainly in the change from a stochastic field to a mean one, but in the

Critical Behavior of Nonideal Systems. Dmitry Yu. Ivanov
Copyright © 2008 WILEY-VCH Verlag GmbH & Co. KGaA, Weinheim
ISBN: 978-3-527-40658-6

suggestion about the analytic properties of a certain "free energy functional." This assumption, in fact, implies the neglect of fluctuations ([51], p. 47).

Mentioned neglect of fluctuations is the reason why, as had already become clear by the middle of the 1960s,[1] experiment does not agree quantitatively with the conclusions of "mean-field" type theories. Even if this is so, this does not stop them being universal zeroth-order approximations. The main feature which distinguishes the modern approach to the description of critical phenomena from any mean-field-type theory is that it takes into account the role of anomalous growth of fluctuations, correlated on all scales from the microscopic distance between molecules up to the correlation length, ξ, which is the maximum spatial dimension for the given temperature [61, 62].

In one-component liquids, order-parameter fluctuations are density fluctuations, which, in turn, lead to the appearance of fluctuations of the specific entropy (see Section 4.3). In liquid mixtures close to the mixing critical point, concentration is the order parameter. The behavior of the radial distribution function and the correlation length in both pure liquids and mixtures is totally analogous. However, in the latter case, the role of the "droplet," performing a Brownian motion, is played by concentration fluctuations, and as a consequence, the reciprocal diffusion coefficient of liquid mixtures close to the mixing critical point behaves like the thermal diffusivity of one-component liquids close to the liquid–vapor critical point. The viscosity of mixtures has the same weak anomaly as in one-component systems, but thermal conductivity remains finite [270].

Moreover, experiment shows [93, 271–273] that, in contrast to pure liquids, in the description of binary mixtures [63,64] the role of nonanalytic corrections, which have already been discussed in application to the analysis of data of pure liquids (see Chapters 1 and 2), is negligible, if they exist at all. The typical situation for liquid mixtures in the neighborhood of the liquid–vapor critical points line (so-called *plait points*) is more complex. This is because we are dealing here with two order parameters: density and concentration. Therefore, it seems natural that, far from the critical point, density fluctuations will dominate, stimulating the singularity of thermal conductivity in the same way as for one-component liquids, while close to the critical point, concentration fluctuations guarantee a finite value for thermal conductivity.

However, experiment did not confirm these expectations. In the liquid helium isotope mixture ^{3}He–^{4}He [274, 275] and in the binary mixture methane–ethane [276], clear features of the critical behavior of thermal conductivity, characteristic for one-component liquids, were found. At the same time, the expected crossover to finite values of thermal conductivity was not detected (see, e.g., [277] and references therein). Here we do not go any further into details on what happens to the kinetic coefficients in critical mixtures (for more details, see [278, 279]), as

---

1) It is well known that which is new is something old which has been well forgotten. As was already mentioned (see Chapter 1), J. E. Verschaffelt at the end of the 19th century, using his own and experimental facts cited in the literature, already reached in his estimates almost modern values of critical indices for pure liquids [46].

this lies outside the scope of this book. However, it is worth emphasizing once more that despite the fact that pure liquids and their mixtures belong to the same universality class, the universality of their dynamic critical behavior has a much more restricted character than when static properties are discussed. This statement implies that in systems which formally belong to the same universality class and which demonstrate the same static behavior, completely different dynamic effects can be observed ([52], p. 266).

There is a simple explanation for this. Having the same static action (more precisely, functional action) you can have many different dynamic models with different intermodal connections ([51], p. 565). Berg and Gruner's paper is a quite remarkable illustration of this fact [280]. In it a surprisingly low value for the critical index of viscosity ($x_{\overline{\eta}} = 0.044$) for one of the polymer solutions compared to pure liquids and mixtures ($x_{\overline{\eta}} = 0.054$ or 0.065) was found; more details on the critical index of viscosity are given in Section 4.4. At the same time no differences from the latter, as a rule, are found in the static critical behavior of polymer solutions (see, e.g., [281]).

Finally, one more feature of critical phenomena dynamics should be mentioned. Its formation and development was and remains closely related to light scattering as light is a unique instrument for investigating such kinds of effects, and the spectral distribution of light scattering contains information on correlation functions and fluctuation kinetics. The results of experiments on light scattering were the source of many ideas and approaches to critical dynamics, and they made it possible to check and correct the conclusions of different theories developed for it. We are not going to tear apart their natural connection and so let us first remember some facts about light scattering.

4.2
Critical Fluctuations: Light Scattering Intensity

The differential cross section of scattering is defined as the scattering coefficient or Rayleigh's constant

$$R_{sc} = \frac{1}{V}\frac{d\sigma}{d\Omega} = \frac{I(\theta, \psi)r^2}{I_0 V}, \tag{4.1}$$

where $I(\theta, \psi)$ is the mean intensity of light falling into the solid angle, $d\Omega$, located at the observation point with the polar coordinates r, θ, and ψ: $d\Omega = \sin\theta d\theta d\psi$, I_0 and V are the intensity of the incident light and scattering volume, respectively. John W. Rayleigh (Strutt)[2] suggested that the incoherence of secondary waves, radiated by gas molecules, is connected with the independence of their thermal motion [284] (here we are not using Rayleigh's formula itself as it is inapplicable for liquids, not even approximately applicable [285]). Later, in

---

[2] J. Rayleigh (1842–1919): Nobel Laureate for Physics (1904) for his investigations of the densities of gases and the discovery of argon in connection with these studies.

1907, Mandelstam [286] showed that when there are lot of scattering particles their thermal motion does not violate the wave coherence (Rayleigh's suggestion is valid for only a small number of particles [286]). Therefore, despite Rayleigh's opinion, molecular motion cannot be the reason for light scattering. Nevertheless, Rayleigh's formula is correct, although its justification should be different.

It has been known since Fresnel that if light passes through a completely homogeneous medium, then, as a result of interference, all the waves, scattered from every point of the medium, will be extinguished in all directions apart from the forward direction. Smoluchowski [35] was the first to notice that real liquids are never completely homogeneous due to their density fluctuation. Then, Albert Einstein[3] attempted a quantitative realization of this idea. Based, in accordance with Boltzmann's statistics, on the additivity of entropy of different parts of a given volume of the system under investigation, he developed [287] the thermodynamic theory of light scattering in liquids and obtained a formula connecting the scattering intensity with physical constants of the respective media:

$$R_{sc}(q=0) = R_{sc}(0) = \frac{k_0^4}{16\pi^2} \sin^2 \Phi \left(\rho \frac{\partial \varepsilon}{\partial \rho} \right)_T^2 k_B T K_T. \tag{4.2}$$

Here q is the scattering wave vector; $k_0 = 2\pi/\lambda$ is the wave number (λ the wavelength of incident light in vacuum); Φ is the angle between the polarization direction of planarly polarized light and the direction of observation, $\vec{r}$; ρ and ε are density and dielectric constant, understood as the square of the refractive index ($\varepsilon = n^2$), respectively; and k_B is the Boltzmann constant. The scattering wave vector $\vec{q} = \vec{k}_s - \vec{k}_i$, taking into account the approximate equation $\left| \vec{k}_i \right| \approx \left| \vec{k}_s \right| = k = (2\pi n/\lambda)$, is determined as

$$q = 2k \sin \left(\frac{\theta}{2} \right) = \frac{4\pi n}{\lambda} \sin \left(\frac{\theta}{2} \right), \tag{4.3}$$

where $\vec{k}_i$ is the wave vector of the incident radiation, $\vec{k}_s$ is the wave vector of scattered radiation, n is the refractive index of scattering media, and θ the scattering angle. For gases, Einstein's formula agrees with Rayleigh's formula [288]. Taking into account the divergence of isothermal compressibility, Einstein's formula could even explain critical opalescence in one-component systems with noncorrelated fluctuations. Of course, for calculating the angular dependence for strong scattering near the critical point, Eq. (4.2) is inconsistent. However, for forward scattering ($q = 0$) it remains true even in the critical region.

In a next step, Ornstein and Zernike, who did not accept the idea about entropy additivity of different parts of the volume, in fact turned to the mutual correlation of density fluctuations. Taking into account the connection which exists between susceptibility and correlation functions (see, e.g., [41]), they were the first, in

---

3) A. Einstein (1879–1955): Nobel Prize Laureate in Physics (1921) for important mathematical physics investigations, especially for his explanation of the photoelectric effect.

the development of Einstein's approach, to work out a calculation scheme [36] for large-scale fluctuations and obtained the now well-known formula for light scattering, whose intensity strongly depends on the scattering wave vector

$$R_{sc}(q) = \frac{R_{sc}(0)}{(1 + q^2 \xi^2)}. \tag{4.4}$$

Here ξ is the correlation length which is connected with the isothermal compressibility by the relation

$$\xi^2 = \xi_0^2 \left(\frac{K_T}{K_{0T}} \right) = \xi_0^2 \tau^{-\gamma} = \xi_0^2 \tau^{-2\nu}, \tag{4.5}$$

where ξ_0 is the so-called direct correlation length, a nonuniversal constant, depending on intermolecular potentials, and $K_{0T} = (\rho k_B T)^{-1}$ the isothermal compressibility of ideal gas. Equation (4.4) is a consequence of the fact that with high values of r the spatial correlation function $G(r)$ in the structure factor

$$S(q) = \left[1 + \rho \int_V \exp(i\vec{q}\vec{r}) G(r) d\vec{r} \right] \tag{4.6}$$

has the following form (see, e.g., [289]):

$$G(r) = \frac{1}{4\pi\rho\xi_0^2} \frac{\exp(-r/\xi)}{r}. \tag{4.7}$$

It should be noted that although the Ornstein–Zernike function can be regarded as only a natural generalization of the van der Waals theory, i.e., a classical theory [6], it was and, on the whole, still is very successful in describing a variety of experimental data. In addition, this approximation is still often used as a starting point for various attempts at creating more complex theoretical approaches.

For example Fisher, based on the exact result $G(r) \sim r^{-1/4}$ well known for a two-dimensional Ising model, proposed [57, 101] a modified form of the Ornstein–Zernike correlation function

$$G(r) = \frac{1}{4\pi\rho\xi_0^2} \frac{\exp(-r/\xi)}{r^{d-2+\eta}}. \tag{4.8}$$

Fluctuation theory of phase transitions shows that the difference between Eqs. (4.8) and (4.7) is not that large if one uses the exact, but not calculated in the Landau approach, value for the correlation radius, ξ ([109], p. 471). The value of the index η is small and for three-dimensional Ising models, depending on the method of calculation, is found within the limits $0.031 \le \eta \le 0.041$ (see, e.g., [40]). Therefore, it is not very easy to discover it experimentally, if at all (see Section 4.4 for more details). As a result, for Rayleigh's constant we get

$$R_{sc}(q) = \frac{k_0^4}{16\pi^2} \sin^2 \Phi \left(\rho \frac{\partial \varepsilon}{\partial \rho} \right)_T^2 k_B T K_T \left(1 + q^2 \xi^2 \right)^{(\eta/2)-1}. \tag{4.9}$$

With Fisher's corrections, Eq. (4.9) predicts a slightly reduced curvature of the dependence of the inverse scattering intensity on q^2.

In turn, the full scattering cross section h, known as the extinction coefficient, can be calculated as an integrated critical part of scattering from Eqs. (4.4) or (4.9) on the overall solid angle:

$$h = \int_\Omega R_{sc}(q) d\Omega(\theta, \psi)$$

$$= \frac{\pi^2}{\lambda^4} k_B T \left(\rho \frac{\partial \varepsilon}{\partial \rho} \right)_T^2 \pi K_T \left[\frac{2w^2 + 2w + 1}{w^3} \ln(1 + 2w) - \frac{2(1 + w)}{w^2} \right]. \tag{4.10}$$

Here, $w = 2(k\xi)^2$ holds. Equation (4.10), which first appeared in [195], suggests as a condition of integration that $\eta = 0$ and the perpendicularity of the electrical vector of planar polarized light of plane observation. This result can be achieved as a special case of a more general calculation (see Eq. (6.39)). For small values of w, retaining the three first terms of the series in the expansion of the function $\ln(1 + w)$, we get

$$h^{(0)} = \frac{8}{3} \pi \left[\frac{\pi^2}{\lambda^4} k_B T \left(\rho \frac{\partial \varepsilon}{\partial \rho} \right)_T^2 \right] K_T \equiv \frac{8}{3} \pi A K_T, \tag{4.11}$$

where

$$A \equiv \frac{\pi^2}{\lambda^4} k_B T \left(\rho \frac{\partial \varepsilon}{\partial \rho} \right)_T^2. \tag{4.12}$$

By combining Eqs. (4.10) and (4.12), we obtain

$$h = \pi A K_T \left[\frac{2w^2 + 2w + 1}{w^3} \ln(1 + 2w) - \frac{2(1 + w)}{w^2} \right]. \tag{4.13}$$

Passing from w to the more natural variable $k\xi \equiv x_0$ we get

$$h = \frac{\pi A K_T}{2x_0^2} \left[\left(2 + \frac{1}{x_0^2} + \frac{1}{4x_0^4} \right) \ln \left(1 + 4x_0^2 \right) - \left(2 + \frac{1}{x_0^2} \right) \right]. \tag{4.14}$$

It is known that in the absence of absorption the intensity of transmitted light I_{tr} is connected with h in accordance with the Bouguer–Lambert–Beer law (see, e.g., [290], p. 232; [291], p. 505) as

$$h = \frac{1}{l_s} \ln \left(\frac{I_0}{I_{tr}} \right), \tag{4.15}$$

where l_s is the length of light's path in the scattering medium. It is Eq. (4.15) which is commonly applied to get experimental values of extinctions, whose measurements are widely used [93, 292]. These experiments allow us to determine the critical index of compressibility γ, the amplitude ξ_0, and the critical index ν of the correlation radius. These questions are discussed further in Chapter 7.

4.3
Kinetics of Critical Fluctuations: Light Scattering Spectrum

As the fluctuations, which are responsible for light scattering, do not stay unchanged with time, the spectral composition of scattered light differs from the incident one. To describe the temporal behavior of fluctuations one can, in principle, use different pairs of variables, for example, (T, ρ) or (S, p). However, as Landau and Placzek first showed [293], for simple liquids it is entropy and pressure that are independent variables. Therefore, if we consider the density of liquid, ρ, as a function of pressure and entropy, then the fluctuation $\delta\rho$ can be expressed as

$$\delta\rho = \left(\frac{\partial\rho}{\partial p}\right)_S \delta p + \left(\frac{\partial\rho}{\partial S}\right)_p \delta S. \tag{4.16}$$

The thermal motion of molecules in the liquid, like in crystals, also brings about the appearance of sound waves (phonons, or adiabatic pressure fluctuations). Light scattering on these waves leads to the appearance of the Mandelstam–Brillouin duplicate in the spectrum, symmetrically positioned relative to the frequency of incident radiation. Isobaric entropy fluctuations, in contrast, scatter slowly due to thermal conductivity. They are just responsible for the presence of unshifted Rayleigh component in the scattered light spectrum [292]. As a result of the scattering on both types of fluctuations there arises a fine structure of the scattered light spectrum, consisting of three separate lines, the Mandelstam–Brillouin duplicate and the central unshifted Rayleigh component. The fine structure of the spectrum was first experimentally discovered by Gross [294] (highly interesting information on the dramatic history of this discovery is given in the book [295] and in the articles [296, 297]).

The scattering processes on fluctuations of pressure and entropy are noncoherent, while the ratio between the intensities of the central and the shifted components is determined by the ratio between isobaric and isochoric heat capacity, corresponding to the well-known Landau–Placzek formula [292]. Thirty years later, Peter Debye[4] used Landau's idea to describe concentration fluctuations in binary mixtures [298]. For nonpropagating critical fluctuations, the power spectrum $I(q\omega)$, which is Lorentzian with a half-width at half-height equal to Γ, following the Wiener–Khinchin theorem ([290], p. 280), is conjugated, in the sense of a Fourier transformation, with a time autocorrelation function, which is represented, in this case, by a decaying exponential function of the type $\exp(-\Gamma t)$.

In the hydrodynamic range ($q\xi \ll 1$), the Landau–Placzek [293] and Debye equations [298] predict that the width of the Rayleigh line, which in one-component liquids or binary mixtures is, in essence, equal to the rate of decay of the order-parameter fluctuations, can be expressed (for wave vector of scattering q) in the following way:

$$\Gamma = D_T q^2 \quad \text{for simple liquids,} \tag{4.17}$$

4) P. Debye (1884–1966): Nobel Prize Laureate in Chemistry (1936) for investigation of dipole moments and contributions into the studying of molecular structures.

$$\Gamma = Dq^2 \qquad \text{for binary mixtures,} \tag{4.18}$$

where D_T is the thermal diffusivity coefficient and D the diffusion coefficient for binary mixtures.

Equations (4.17) and (4.18) are a particular case of the more general expression (4.19) [299, 300]

$$\Gamma = \left(\frac{L}{\chi}\right) q^2, \tag{4.19}$$

where L is the Onsager kinetic coefficient, and χ is the generalized susceptibility. Approaching the critical point leads to a more rapid growth of the susceptibility than a possible increase in the corresponding kinetic coefficient. This result explains the well-known fact that much more time is needed to establish thermodynamic equilibrium in the system under these conditions.

Taking into account large-scale correlations of typical length ξ, and assuming that the value of ξ does not change with transition from static to dynamic experiments, the Landau–Placzek–Debye theory can also be used in the transition area (at $q\xi \leq 1$), in the refined version including Fixman–Botch corrections [301]

$$\Gamma = Dq^2 \left(1 + q^2\xi^2\right)^{1-(\eta/2)}. \tag{4.20}$$

Further, we shall briefly look at three well-known dynamic theories which attempt to describe the peculiarities of critical fluctuation dynamics: the uncoupled mode theory (UMT) [302, 303], the dynamic droplet model [304, 305], and the mode coupling theory (MCT) [306–309]. All these theories originate from very similar, but nevertheless different, premises.

Ferrell, in the calculation of the diffusion coefficient using Kubo's formulas (see, e.g., [290], p. 539; [310], p. 532), broke up the quadruple correlation function into product of binary correlators of concentration (equilibrium part) and velocity (nonequilibrium part) [302]. By doing this, because of oddness, cross value correlators drop out. Ferrell interpreted this fact as the absence of interaction between diffusion and viscous modes, which is where the name, the theory of uncoupled modes, comes from. The expression obtained for the rate of decay, within the scope of this theory [302, 303], differs only slightly from Eq. (4.23) which Kawasaki obtained within the limits of MCT. Looking ahead, it should be said that, subsequently Kawasaki, apparently quite correctly, called UMT a simplified version of his theory.

Another version of the theory of critical fluctuations is the dynamic droplet model, put forward in [304, 305]. The authors of these works started from the fact that the q-dependence of half-width of the Rayleigh profile near the critical point can be correctly described if critical order-parameter fluctuations represent clusters (droplets) which diffuse like Brownian particles in a medium with regular viscosity η_S^0. The expression obtained for Γ_c within the limits of the dynamic droplet model looks like

$$\Gamma_c = \frac{k_B T}{6\pi\bar{\gamma}\eta_S^0\xi} q^2 \left[1 + \left(q\xi\right)^2\right]^{1/2}, \tag{4.21}$$

where $\bar{\gamma}$ is a factor close to unity. It is an adjustable parameter, defining the difference between time decay of critical fluctuations from the exponential law. Obviously, the authors of this theory started from the fact that, in all the systems they examined, the experimentally observed shape of the Rayleigh profile had, in fact, a non-Lorentzian form.

According to [304, 305], $\bar{\gamma} \geq 1$ holds, which gives a smaller half-width compared to Kawasaki's theory. Generally speaking, this factor should rather be less than unity, then larger than it, since as it is known if a droplet moves in a liquid with the same viscosity then 6π in Stokes' formula is replaced by 5π ([312], p. 100). At the same time, the droplet model seems to be the simplest and most physical, as, finally, the results of all other approaches can also be interpreted in the spirit of diffusion of the droplets with radius, ξ.

Kawasaki suggested to solve the corresponding hydrodynamic equations within the framework of the so-called mode coupling approach [306]. Initially expressed by Fixman [313], this idea was further developed by Kawasaki [306–308] and Kadanoff and Swift [309]. The fundamental idea of the mode coupling theory lies in the fact that kinetic modes (thermal flow, viscous flow and pressure waves) not only interact with each other but interact in a nonlinear way. Sound waves (pressure waves) can convert into thermal and viscous modes. Viscous modes, in turn, can also fall into two, three, or four thermal modes. By using a semi-microscopic calculation, Kadanoff and Swift discovered that intermediate states, which include one mode each of viscous and thermal flows, give the predominant contribution to the anomalous behavior of thermal conductivity. In estimating the relaxation rate of these decay processes they were able, in particular, to show that the thermal diffusivity (D_T) of pure liquids tends to zero, just like the inverse correlation length, i.e., with the critical exponent, equal to v,

$$D_T = \frac{\lambda}{\rho c_p} \sim \tau^{\psi} \sim \tau^{v}, \qquad (4.22)$$

where λ is the thermal conductivity coefficient and c_p is the specific heat capacity under constant pressure. It is known that c_p behaves just like K_T close to the critical point, so the critical index of thermal diffusivity, ψ, can be represented as the difference $\gamma - \phi$ (γ and ϕ are critical indices of isothermal compressibility and thermal conductivity, respectively). Therefore, it can be concluded that the value of the critical index of thermal conductivity is $\phi = \gamma - \psi$, or $\phi = \gamma - v$.

Later Kawasaki [307], generalizing Kadanoff's and Swift's calculations, suggested an exact integral expression for the decay rate (Γ_c) of fluctuations near the critical point (i.e., in fact for the half-width of the Rayleigh line). Using the Ornstein–Zernike correlation function to estimate the integral, Kawasaki obtained the now well-known formula (4.23):

$$\Gamma_c = \frac{k_B T}{6\pi\bar{\eta}^*\xi} q^2 \Omega_K (q\xi), \qquad (4.23)$$

where

$$\xi = \xi_0 \tau^{-v}, \quad \tau = \frac{T - T_c}{T_c}, \qquad (4.24)$$

and $\overline{\eta}^*$ is the "high-frequency" shear viscosity, the effective weighted mean of all these viscous modes, and $\Omega_K(q\xi)$ is the Kawasaki function

$$\Omega_K(x) = \frac{3}{4x^2}\left[1 + x^2 + \left(x^3 - \frac{1}{x}\right)\text{arctg}x\right]. \tag{4.25}$$

For small values of x, the Kawasaki function leads to the following expression:

$$\Omega_K(x) \approx 1 + \frac{3}{5}x^2 - \frac{1}{7}x^4, \tag{4.26}$$

and for large values of x, it reads

$$\Omega_K(x) \approx \frac{3\pi}{8}x + \frac{3}{4x^2} - \frac{3\pi}{8x^3}. \tag{4.27}$$

It has to be mentioned that in an earlier experimental paper [314] the dependence $\Gamma \sim q^3$, which differed significantly from the Landau–Placzek–Debye–Fixman formula, $\Gamma \sim q^2$, was first detected.

It is easy to see that in the hydrodynamic regime from Eq. (4.23), taking into account Eq. (4.26), one obtains a formula of the type of Eqs. (4.17) and (4.18). For intermediate cases, which, following Bergé and colleagues [314–316], we can call nonlocal hydrodynamic regimes ($x \leq 1$), the only difference in Eq. (4.23), taking into account Eq. (4.26), from the Fixman expression (4.20) is the presence of the multiplier $3/5$ at x^2 (in Eq. (4.23) $\eta = 0$ is assumed (see below for more details)). For large values of x, i.e., in a nonlocal critical regime, the same formula (4.23) is fundamentally different,

$$\Gamma_c = \frac{k_B T}{16\overline{\eta}^*}q^3, \tag{4.28}$$

which confirms the result first obtained in [314].

To derive his formula, Kawasaki used a correlation function of Ornstein–Zernike type, Eq. (4.7), and started from the fact that the viscosity $\overline{\eta}$ does not have any anomalies in the critical point. It should also be noted that, in this initial version of Kawasaki's formula, the value that was used as a viscosity coefficient did not have a very clear physical meaning. To develop the theory further, Kawasaki and Lo [317] took into account the frequency dependence of viscosity, its nonlocal character, and also found the connection between high-frequency ($\overline{\eta}^*$) and the normal ($\overline{\eta}$) macroscopic shear viscosities (see also [318] and [319], p. 356). As a result, Eq. (4.23) took on its modern appearance

$$\Gamma_c = R\frac{k_B T}{6\pi\overline{\eta}\xi}q^2\Omega(x). \tag{4.29}$$

Now, Eq. (4.28) can be represented in a more general form as

$$\Gamma_c = R\frac{k_B T}{16\overline{\eta}}q^z, \tag{4.30}$$

where z is the new dynamic critical index [283], R is the universal dynamic amplitude, and $\Omega(x)$ is the dynamic scaling function from the scaled variable $x \equiv q\xi$.

4.4
Dynamic Critical Indices and Universal Amplitude

The asymptotic behavior of $\Omega(x)$ is such [283] that

$$
\Omega(x) = \begin{cases} 1 & \text{for} \quad x \ll 1 \\ \sim x^y & \text{for} \quad x \gg 1 \end{cases},
\tag{4.31}
$$

where $y = 1 + x_{\overline{\eta}}$, and $x_{\overline{\eta}}$ is the critical index of viscosity. Comparing Eqs. (4.30) and (4.31) we can see that $z = 3 + x_{\overline{\eta}}$. The renormalization group (RG) theory results in a value of z slightly larger than 3, i.e., $z = 3.065 \pm 0.003$. It is known (see, e.g., [283, 307]) that the shear viscosity $\overline{\eta}$ diverges, although weakly, at the critical point according to the law

$$
\overline{\eta} = \overline{\eta}_0 \left(\frac{\xi}{\xi_0} \right)^{x_{\overline{\eta}}} = \overline{\eta}_0 \left(\tau^{-y_{\overline{\eta}}} \right).
\tag{4.32}
$$

Here $x_{\overline{\eta}}$ and $y_{\overline{\eta}}$ are the critical indices of viscosity in these two different expressions. Therefore, it is clear that $y_{\overline{\eta}} = v x_{\overline{\eta}}$. When it is said that the viscosity anomaly is multiplicative, it is meant that the regular part $\overline{\eta}_0$ is not added to the "critical" one as usual, but multiplies with it. Experimental evidence showing that viscosity anomalies should not be disregarded when describing the scattering spectrum was first obtained in [315, 320]. The critical index $x_{\overline{\eta}}$ (therefore, also $y_{\overline{\eta}}$) can have different values depending on the method of calculation. In particular, within the framework of MCT and UMT [307, 321], one obtains $x_{\overline{\eta}} = 8/(15\pi^2) \cong 0.054$, but when the RG approach is used [283, 322] $x_{\overline{\eta}} = 0.065$.

Within the framework of RG, the well-known so-called H and H' models give a good impression on the critical dynamics of simple liquids and binary mixtures [283, 322]. With their help, critical indices which characterize the singularity of transport coefficients, as well as the corresponding scaling functions, could be calculated quite precisely. The paper [322] was the first to give a correct RG description of the critical behavior of shear viscosity and to arrive at the value 0.065 for the viscosity index, instead of 0.054 in the case of MCT. Despite this difference, the result of [322] can, nevertheless, be considered as a RG confirmation of MCT [272] in the suggestion that $\gamma = 2v$.

As for the universal dynamic amplitude R, the simplified analysis within the framework of Kawasaki mode coupling theory gives its value $R = 1$ [302, 307, 308]. At the same time, in Kawasaki's and Lo's paper the value of R was slightly higher than 1 [317]. Subsequently, the value of R was also calculated [322] to first order in ε within the framework of the "H'-model" by means of the RG-approach to critical dynamics. Two values were obtained in this paper: $R = 0.79$ and 1.20. The authors considered the latter value to be more realistic. Although it was already mentioned on a different occasion, let us note that the value 6π, which appears in the famous Stokes–Einstein formula for the diffusion coefficient, is in fact replaced by 5π $(R/(6\pi) = 1/(5\pi))$, which corresponds to Stokes' law for spherical droplets moving in media with the same viscosity as the liquid of the droplet itself ([312], p. 100).

The authors of [321], in contrast, consider the values $R = 1.20$ and $x_{\bar{\eta}} = 0.065$ to be incorrectly enlarged and follow a calculation which, allowing than to avoid the ε-expansion, once more leads to the previous (MCT) value for these quantities, 1.04 and 0.054, respectively.

Another attempt to define these numbers more precisely within the framework of "*H*-models" using the renormalization group approach, but this time taking into account viscosity anomalies, was performed by Paladin and Peliti [323]. This allowed them to propose instead of the Kawasaki function a new universal function

$$\Omega_P(x) = \left[\Omega_K(x)\right]^{1-\eta-x_{\bar{\eta}}} \left(x^2 + 1\right)^{(\eta/2)+x_{\bar{\eta}}}. \tag{4.33}$$

In the original paper [323] this function was used for $\eta = 0$. In the form of Eq. (4.33) it, apparently, first appeared in [324]. As for the values under discussion, when using Eq. (4.33) within the framework of a single-loop approximation where $\eta = 0$ (the RG technique, as detailed and clearly as it is at all possible for such a complex subject, is set forth in Vasil'iev's book [51]), they appeared to be equal to 1.075 and 0.07, respectively [323]. However, the comparison with the experiment carried out in this work shows that the combination 1.16 and 0.065 does not describe the experimental data worse.

Actually, this result should not be considered as a great surprise. Indeed, the difference between the suggested modifications is not large and appears only in significantly nonhydrodynamic regions where ($q\xi \geq 3-5$), i.e., in the direct vicinity of the critical point, where the precision of the experiment, for obvious reasons, noticeably decreases. Apart from this, and here is the greatest difficulty, not only the critical but also the regular part has to be included into the consideration [317, 325, 326]. It is these two circumstances which, more often than not, explain the difference between, on the one hand, theory and experiment, and on the other hand, differences between different experiments [93, 272, 273, 326–329].

Thus, for example, Beysens discovered [93] that the experimental data for seven different mixtures show that $R = 1.16 \pm 0.01$, while a lower value is given in [328]: $R = 1.01 \pm 0.04$. It is also noted there that the authors, while working out their own experimental data on dynamic light scattering, were unable to establish which of the two values of the critical index of viscosity, 0.054 or 0.065, should be given preference. And, finally, in his later investigations, where more exact experimental data were obtained, Beysens and his colleagues concluded that, after all, $R = 1.06 \pm 0.06$ [324, 330].

From all this, we can only conclude that the strong dependence of the result on the chosen method of data processing does not allow one to, definitely, assign definite values to the universal dynamic amplitude R and the critical index of viscosity, $x_{\bar{\eta}}$. Also, in accordance with the results of the papers [280, 281], one of which was already mentioned at the beginning of the chapter in connection with the surprisingly low value of the critical index of viscosity $x_{\bar{\eta}}$ for polymer solutions, the correct model for describing viscosity should take into account the influence of gravity (pure liquids) and an adequately chosen crossover function. This is, essentially, what we were talking about above.

To round off, it should be mentioned that, as for the dynamic scaling function, apart from Eq. (4.33), other of its forms are used (see, e.g., [281]), which satisfy the conditions (4.31), in particular that proposed in [326],

$$\Omega_B(x) = \Omega_K(x) \left(1 + b^2 x^2\right)^{x_{\overline{\eta}}/2},$$ (4.34)

where with respect to the coefficient $b = 0.55$, although obtained by theoretical means, it has to be noted that its meaning and value are not completely unambiguous. Function (4.33) proposed by Paladin and Peliti, due to its lack of adjustable parameters, and also because of the great simplicity and clarity of its derivation, seems to us to be, physically, more substantiated.

Finally, let us investigate the situation concerning the experimental determination of one more dynamic index, the index η. This index was first introduced by Fisher in the above-mentioned article [57], which, according to Stanley ([26], p. 159), is one of the best overviews of the original work of Ornstein–Zernike [36] and the discussion connected with it. Since that time this index has remained the only important theoretical prediction which still has not been precisely confirmed experimentally. Even Fisher, many years after the "birth" of this exponent, called it "elusive" [331]. This characteristic is confirmed by the real situation. Let us look at this in more detail. Experimental results before 1975 were systematized in [332] and, later, in [333]. The authors of the first work stated that "no experiments to date allow us to directly and unambiguously establish that the critical index $\eta > 0$." A classical example of works on this theme is the very detailed and thorough investigation [334]. The authors of this work reasonably suggested that to determine the value of the small index η from a combination of large ones [57] via

$$\gamma = (2 - \eta)\, v,$$ (4.35)

would not be possible with sufficiently high precision. They attempted to determine it "directly," comparing the experimental values of the intensity of scattered light close to the critical demixing point for weakly opalescent mixture 3-methylpentane–nitroethane ($\Delta n = 0.0129$) to the theoretical ones. Such a small difference in refractive indices allowed them to get close to the critical point to 10^{-6} with respect to relative temperature.

Multiple scattering corrections were made employing the double scattering approximation, as in such a critical point vicinity the contribution of double scattering even in such a weakly opalescing system can reach up to several percent [335]. While carrying out a statistical analysis of their experimental data the authors attempted to find among the set of correlation functions the one for which the description was more reliable. As this paper [334] is widely known and appeared in all reviews, we will only draw the attention to those details which, up to now, have not had the spotlight put on them. It turned out that the choice of one or another correlation function does not affect the precision of the description of the experimental data. It remains within the limits of 0.4–0.5%, while the value of the critical index η changes within very wide limits and errors in the determination reach up to 100%.

Further analysis of the results of the work [334], carried out in [141], showed that there is a noticeable correlation between the values of η and the choice, not of one or other correlation function, as the authors of [334] suggested, but of one or another value of the direct correlation length, ξ_0. It was discovered that even a 10–15% variation of ξ_0 leads to "catastrophic" changes in the values of η. As it is too much to hope for a significant increase in the accuracy of determination of ξ_0, whose spread of values, by different authors, for one and the same substance reaches 30–40%, one can conclude that the experimental determination of Fisher's critical index using the suggested methods does not hold much promise. We need to look for new approaches.

4.5
Scattering of Higher Orders

As we get closer to the critical point, very high multiplicity scattering starts to contribute more to the anomalous growth of scattering intensity. These processes can be neglected far from the critical point [336]. Close to the critical point, it is natural and convenient to divide the structure of the scattered light into single, double or multiple components. This is due to the fact, that depending on the system under investigation and its degree of vicinity to the critical state, two extreme cases can be realized, single and multiple scattering. Double scattering will play the role of either a severely distorted factor for gently opalescent systems, or of a transitional process for strongly opalescent ones.

There exist a variety of serious studies [337–348], reviews [349–352], and books [353–356] dedicated to the general solution of the problem of electromagnetic wave scattering of arbitrary multiplicity, including states close to the critical point. However, as it is not always easy to use these general theoretical results, it was attempted to avoid the problem of taking into account multiple scattering by decreasing its influence or reducing its role to small corrections. It seemed that this aim could be achieved either by decreasing the thickness of the investigated material layer, as it was done, for example, in [357] and [358] where cuvettes with a thickness of 200 μm and of 13 to 0.5 μm were used respectively, or by choosing components for binary mixtures with close refractive indices [326, 334, 359]. However these measures, which only partly give the desired result, bring about additional difficulties. For example, in too thin cuvettes, because of the commensurability of thickness of the layer of material and typical fluctuation size, there could possibly even be a change in the type of critical behavior [205, 360–363] (see also Chapter 2). In the case of components in mixtures with close refractive indices, because of the weakness of scattering, the possibility of carrying out qualitative spectral measurements far from the critical point becomes problematic. These measurements are necessary for correctly accounting for nonsingular, background contributions into the spectrum half-width [319, 320, 326]. Moreover, corrections, connected with at least double scattering [364], have to be included even in these cases. Therefore, it can be stated that multiple scattering, arising as a result of the fast growth of the order-parameter fluctuations close to T_c, and therefore of the growth of the extinction, should be attributed, in fact, to unavoidable factors in the studies of critical phenomena using optical methods.

There are a lot of papers devoted to multiple scattering (see, e.g., the reviews [289, 292, 349–352, 365]). The term "multiple" scattering often implies scattering of any multiplicity higher than the first, even if the average multiplicity is not very high. Without going into excessive detail, as there are so many existing reviews on this topic, we will mention only those works which, in our opinion, played the most vital part in the research of scattering of the highest multiplicity. First of all we would like to note the fundamental work of Shifrin, "Light Scattering in a Turbid Medium" [353]. Despite the fact that it does not directly examine the effects of multiple scattering, this book, by right, could and should have become the basic manual for everybody studying scattering. But, unfortunately, as it was published in 1951, it has already become a bibliographic rarity.

The theoretical examination of double scattering light's integral intensity, first conducted in the works of Chaly [366] and Oxtoby and Gelbart [318, 367, 368], showed its nontrivial dependence on the linear sizes of the scattering volume, smooth dependence on the scattering angle, and sudden growth as it approaches the critical point. A detailed calculation of the scattering intensities of all polarizations was also carried out within the limits of spherical geometry, neglecting any light attenuation due to extinction [318, 367, 368]. Barabanenkov and colleagues [343–345, 349] made an important contribution to the studies of the problems of multiple scattering both on the ensemble of particles and on critical fluctuations in the framework of radiation transport theory. Kuz'min suggested [340] a microscopic theory of multiple light scattering, based on the idea of diagram representation of the resultant field in the form of the field's superposition, resulting in single and multiple reradiation, which is in essence the theory of perturbations on multiple scattering. He then found an important estimate of the parameters of the spectrum of double scattering $\Gamma_2 \gg \Gamma_1$ far from the critical point and $\Gamma_2 \ll \Gamma_1$ close to it. These result was partly confirmed in experiments [369]. In Reith and Swinney's work [370] in the spirit of the ideas outlined in [318, 367, 368] a more realistic, in the experimental sense, cylindrical geometry was analyzed, and the scattering constant was determined by the relation between integral intensities of single and double light scattering. In [371, 372], the cylindrical geometry analysis was continued taking into account the frequency dependence of the order-parameter correlation function.

The pioneering work of St. Petersburg University's physicists [373] should also be mentioned, where for the first time the original method of separation of multiple (double) scattering was suggested, which later became generally employed. The main idea of this method is to exclude the intersection of the volume irradiated by incident light with the volume from where scattered light comes. In their subsequent series of papers, these authors managed (as a result of sensitive measurements of angular dependence of integral intensities of the polarized and depolarized component of double scattered light in the wide neighborhood of the critical point of demixing in binary mixtures in real geometry of the experiment) to correctly account for the contribution of double scattering and using its temperature dependence, to determine critical indices of susceptibility and the correlation radius, and also the value of ξ_0 [374–376].

In Beysen's and Zalczer's works [369, 377], a method of calculating the double scattering spectrum, taking into account extinction, was proposed, and for the first time this type of spectrum was experimentally studied close to the critical point of demixing [369]. The papers [378, 379], where special attention was paid to clarifying the character of statistics for scattering of different multiplicities, became, methodologically, very important. They showed experimentally that, as with scattering close to the critical point [378] so with scattering on latexes [379], independently from the multiplicity of light scattering, the statistics is Gaussian. From general considerations, based on the central limit theorem, according to which the field of N sources has Gaussian statistics, if all sources are independent and their number is high ($N \rightarrow \infty$) ([291], p. 425), it is clear that if in a system under single scattering conditions statistics are Gaussian, then an increase in the number of scattering acts can in no way violate this state. Consequently, the information received as a result of hetero- or homodynation, is equivalent even in the case of multiple scattering (for more details on this topic, see Chapter 5). A general analysis of the spectral structure of double scattered light was conducted in [337], and quantitative calculations taking into account the real geometry in [380]. The influence of gravity on the spectral and integral intensities of double scattered light in the neighborhood of the liquid–vapor critical point was examined in [381]. It should also be noted that in all mentioned works a double scattering was considered as a result of two consecutive independent acts of scattering.

Summarizing those ideas, which had formed by the beginning of the 1980s, it might be thought that multiple scattering makes a large contribution to the intensity of light scattering. However, its influence on the critical opalescence spectrum is small and only leads to the appearance of "non-Lorentzianess" in the form of the spectral line [328, 335, 350]. The spectrum half-width of multiple (Γ_m) and single (Γ) scattering decreases as temperature approaches to the critical [369], and the angular dependence of the spectrum half-width of multiple scattering is missing [350, 382–385]. As studies [386–392] showed later, these conclusions are relevant, mainly, for double scattered light, and there is not enough reason to extrapolate them to high multiplicity scattering.

5

Critical Opalescence: Modeling

5.1
Introduction

Critical opalescence is one of the most fundamental features of critical behavior. The nature of critical opalescence, like the nature of critical phenomena in general, is determined by anomalous growth of order parameter fluctuations on approaching the critical point [40,41,47,55]. Investigations into critical opalescence have always been the subject of great interest [5,7,26,31,100,117,285,289,292]. Light is an ideal instrument for studying critical opalescence, as its influence on the investigated medium is negligible. As the light wavelength close to the critical point is comparable to the typical size of fluctuations, by using light scattering one can obtain information on the behavior of microscopic characteristics of matter, such as the correlation radius of fluctuations, ξ, and their decay time, $\bar{\tau}$ ($\bar{\tau} \sim 1/\Gamma$, where Γ is the half-width of Rayleigh components in the scattering light spectrum).

As the critical point is approached the scattering multiplicity grows, reaching very high values in the case of developed opalescence. Scattered light, naturally, also in this case contains unique information about static and dynamic properties of the scattering media. However to get hold of this information, one must be able to extract it, which is not a simple matter in the case of multiple scattered light spectra [141,350,386–389,393,394]. Therefore, experimentalists investigating critical opalescence usually try different methods to decrease the effect of multiple scattering, and to confine it to small corrections (see, e.g., [357,358]). Theoretically, it is only possible to account for double scattering (see, e.g., [347,375]). It is quite clear, however, that when investigating developed critical opalescence one has to deal with scattering on growing critical fluctuations, that is scattering of really high multiplicity, one cannot get rid of it. So sooner or later its role in this process will have to be closely examined. A large number of investigations have been dedicated to this problem [141,386–395,397,398]. A description and generalization of their results are the subject of the fifth and sixth chapters of this book.

Due to the extreme complexity of this task, the first stage of its solution was the modeling of multiple scattering media using water suspensions of monodisperse polystyrene latexes of different sizes and concentrations. The use of latexes proved

Critical Behavior of Nonideal Systems. Dmitry Yu. Ivanov
Copyright © 2008 WILEY-VCH Verlag GmbH & Co. KGaA, Weinheim
ISBN: 978-3-527-40658-6

to be a simple and rather effective method of achieving a model medium with regulated extinction as the submicron particles of latex have an almost ideal spherical form. The ideal spherical form, apart from anything else, also guarantees the absence of extra contours in the scattered light spectrum connected with particle rotation [336]. In addition, the particles do not interact with each other and, practically, do not absorb in the visible part of the spectrum (for more details, see Section 5.3).

Scattered light correlation spectroscopy [319] occupies a special place among optical methods. It allows us to register the half-width of the central line in the quasielastic scattered light spectrum, with a resolution inaccessible for traditional optical methods. The measured spectrum broadening arises due to light modulation by internal motions of different nature, Brownian motion of particles, suspended in the liquid, dissipation processes of thermal fluctuations, etc. Because of this, correlation spectroscopy gives us a unique opportunity to study the mobility, viscosity, diffusion coefficient, and the thermal conductivity coefficients of liquids using gradientless noninvasive methods. This method is also used to investigate the properties of the medium close to phase transition points (pure liquid, mixtures, liquid crystals, synthetic and biological polymers, including DNA), typical laminar and turbulent flow features, kinetics of chemical reactions, etc., [319, 399–403]. As well as this, correlation spectroscopy can solve a whole range of problems of chemical technology, automation of technological processes, thermal power and food industry (defining the particle size of a highly dispersed media, flow speed), etc.

Correlation spectroscopy has, for a long time, been the routine method of laboratory analysis for single scattering. However, there are many objects, natural and artificial (microemulsions, water-based paints, milk, liquids close to the critical point, etc.), where multiple scattering is prevalent, but transition to single scattering conditions (e.g., by system dilution) for one reason or another is unacceptable. Therefore, the problem of obtaining information from the spectra of quasielastic scattering of high multiplicity is extremely relevant. This task is analyzed in this chapter. It contains the results of investigation of light scattering carried out on model systems of Brownian particles [386, 388, 390, 391, 393, 394] and, essentially, the first world attempts to systematically study the spectra of quasielastic multiple scattering using photon correlation techniques.

5.2
Techniques and Experimental Methods

Correlation spectroscopy of scattered light (quasielastic or dynamic scattering) as a spectral analysis method corresponds to scattering with a small frequency shift due to either the Doppler effect on moving particles, or time evolution of thermal fluctuations which scatter light. At the heart of these methods lies the idea expressed by Gorelik [404] and almost simultaneously by Forrester and colleagues [405] in 1947. However, this method started to be widely used only after the development of lasers, although, in theory "prelaser" sources could be used [406]. A complete and detailed description of the methodical techniques and different applications of correlation spectroscopy can be found, in particular, in Refs. [319, 399–403]. Therefore, we will only look here at the basic features of the experimental setups.

5.2.1
Experimental Setup

5.2.1.1 **General Characteristics**
The experimental setup, whose block scheme is shown in Fig. 5.1, consists of a laser which illuminates the scattering medium, a goniometer with an optical system which forms a light beam, scattered at a determined angle, a photoelectron multiplier (PM), and a multichannel correlator. In addition, this setup includes a photon counter, a thermostat, a frequency-measuring device, and a computer with peripheral devices [391, 407]. Now, with the development and the prevalence of computer technology, correlators are usually made as a separate special card which is placed into a personal computer. As a result, all the equipment becomes compact and even more "intellectual" (see, e.g., [408]).

Scattered light through a system consisting of two diaphragms, a polarizer and, lenses falls on the photocathode of the PM which is photon counting mode operated. Electrical impulses coming out of the PM through an amplifier and discriminator arrive at the entrance of the multichannel correlator. An analogous output of the correlator allows us to observe on the screen (either an oscillograph or a computer monitor) the accumulation of correlation functions which guarantees the possibility of continuous control of the setup's work. The validity of the mathematical treatment can be controlled by the graph showing the deviations

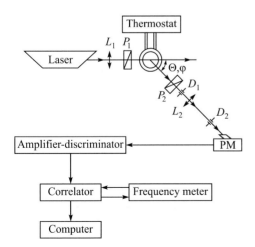

Figure 5.1 Experimental setup for the analysis of light scattering employing the method of correlation spectroscopy.

from the used approximation. By using a frequency meter in the setup, which measures the mean frequency of photocounts, it is possible to keep track of the intensity of the scattered light.

5.2.1.2 The Optical System

The optical system forms a narrow laser beam which illuminates the cuvette with scattering liquid, and also separates a coherent light beam scattered at a certain angle. Then this scattered light is registered by a photon counter system [319]. The laser beam, vertically polarized, goes through the studied liquid placed into the cylindrical glass cuvette normal to its axis. The incident beam is focused by the long focus lens L_1 on the boundary between scattering medium and the cuvette wall. This scattering geometry is very convenient in those cases where the measurements, begun under single scattering conditions, need to be continued on the same setup and with a sharp increase in the scattering medium's extinction. As even the maximum radius of the cuvettes employed (1 cm) is small as compared to the value of f_1, so in the case of single scattering, beam focusing takes place not only on the boundary but also on the axis of the cuvette. Under multiple scattering conditions, it can be supposed that the point light source is located closely to the scattering medium's boundary.

Some peculiarities of the realization of optical experiments in multiple scattering mode should be mentioned. In this case, the direct laser beam does not pass through the scattering medium, whose optical thickness is quite large, but scatters throughout the whole volume. Scattered light, coming out through the surface, spreads nonuniformly over it. A significant part of the light is radiated by the part of the medium's surface which surrounds the incoming laser beam. Undergoing total reflection on the border between glass cuvette and air, this light can fall onto other, less bright, sites of the scattering medium's surface, significantly changing the general distribution of scattered light over the surface. This process can lead to distortions in the measured results. To check whether such distortions take place or not, and if so how real they are, a part of the measurements should be carried out by placing the cuvette in a cylindrical glass vessel filled with immersion medium. As it is usually water suspensions of Brownian particles that are being investigated, water is used as the immersion medium. In most cases, the measuring results in the immersion medium and without it fall within the experiment's margins of error. In the cases where this does not happen, all the measurements should be performed in the immersion medium.

To observe scattering at different angles, a radiation detector (PM) is placed on the rotating base of the goniometer (many common devices only allow one to perform the measurements for only one, fixed (as a rule, $90°$) scattering angle), and the cuvette with the studied liquid is fixed on the axis. The axes of incident and scattered light beams, perpendicular to the cylindrical boundary of the cuvette, form the angle ϕ. Unlike single scattering, in the case of multiple scattering, ϕ is the observation and not scattering angle, θ.

It is well known [319, 399–402] that spectral analysis of light using discussed optical methods requires that the coherence of the beam, incident on the PM, has to be guaranteed. For this reason a spatial filter, consisting of a L_2-lens and D_1 and D_2-diaphragms, is used. The lens is placed in the middle between the cuvette and the PM. The distance between them is equal to a quadruple focal. A circular D_2 diaphragm, whose diameter is a fraction of a millimeter, is placed close to the photocathode which in such cases assigns the required "one coherence area." Close to the lens is a variable aperture diaphragm D_1, which when fully open allows one to make a calibration of the setup. For better vibroprotection, the whole optical part of the setup should be somehow dampened.

A photon counter converts the optical signal into an electrical one which is suitable to be analyzed by a digital correlator (see, e.g., [409]). A photomultiplier is used as the photon detector, and, at the same time, the PM operates at the beginning of the plateau characteristic curve [410] (PM-79, which is the most commonly used photon multiplier, is well known to be suitable for working in a photocount regime [410] (see also [408])). Single electron impulses from the PM's anode convert themselves into a sequence of impulses standardized by amplifier–discriminator used for the suppression of noise impulses whose amplitude is lower than the threshold.

5.2.1.3 **Correlator**

The digital correlator is the basic element of the experimental setup. If one cannot purchase an industrial-made correlator, then the way to choose a specific correlator circuit out of the great variety of circuits, as described in literature, is basically determined by the relative simplicity of preparation and convenience. We can recommend the single bit circuit, similar to that described in [319], which has 60–80 channels. On the one hand, the number of the correlator's channels determines the level of detail of the registered correlation function, and on the other the cost of preparation. As for single, and fortunately, multiple light scattering on Brownian particles and critical fluctuations, the correlation function is quite smooth and steadily decreasing; further growth in the number of channels is fairly pointless (in the carefully carried out analyses [326, 328, 359], insignificant differences in the form of the contour from the Lorentzian for scattering on critical fluctuations were discovered). The minimal sample time, determining the discreteness of the correlation function representation on the time axis, depends on the used chip speed and usually varies within the bandwidth 50–200 ns.

The correlation function of light intensity scattered on Brownian particles or fluctuations in liquid contains, together with an exponentially damped part, also a part independent of time, the so-called pedestal. To separate these contributions, some of the correlator's last channels are usually kept apart from all the other parts of the shift register. This creates a big delay so that the contribution of the correlation of the delayed signal with the current one in these last channels can be disregarded. In these channels, the contributions to the correlation function accumulate, which gives the possibility to calculate the "base line."

In the next part, we will discuss the method of mathematical analysis of the correlation functions.

5.2.1.4 Time Correlation Function for High Scattering Multiplicities

In traditional spectroscopy, radiation is expanded into its spectrum with the help of a dispergating element or monochromatizing filters and then a photodetector measures the power of the spectrum's optical field

$$S(\omega) = \int g^{(1)}(\bar{\tau}) \exp(i\omega\bar{\tau})d\bar{\tau}, \tag{5.1}$$

where

$$g^{(1)}(\bar{\tau}) = \frac{\langle E(t)E^*(t-\bar{\tau})\rangle}{\langle I(t)\rangle}. \tag{5.2}$$

Here $\bar{\tau}$ is the decay time and $g^{(1)}(\bar{\tau})$ is the normalized time correlation function of the first order [319, 411]. It is $g^{(1)}(\bar{\tau})$, which carries on the information equivalent to that contained in the optical spectrum.

The main distinction of correlation spectroscopy lies in the rearrangement of the spectroanalytic element and the photon detector, so that the beam initially transforms in the photon detector's signal, which then undergoes spectral analysis (such a rearrangement of functional elements of the measuring apparatus corresponds to the rearrangement of words in the name of the measured quantity; it is not the power of the spectrum that is measured but the spectrum of power). As the photon detector's signal is proportional to the intensity of light $I(t) = |E(t)|^2$, so the procedure for measuring the spectrum of the power of this signal is equivalent to the measurement of

$$P(\omega) = \int g^{(2)}(\bar{\tau}) \exp(i\omega\bar{\tau})d\bar{\tau}, \tag{5.3}$$

which is the Fourier transform of the time correlation function, not of the first but of the second order

$$g^{(2)}(\bar{\tau}) = \frac{\langle I(t)I(t-\bar{\tau})\rangle}{\langle I^2(t)\rangle}. \tag{5.4}$$

The digital correlator handles the signal which appears as a temporal sequence of photocounts. The result of such a signal's processing is the accumulation of the set of values of the time signal's correlation function, which can be represented as

$$g^{(2)}(\bar{\tau}) = \frac{1}{t_0\langle I^2(t)\rangle} \int_{-t_0/2}^{+t_0/2} I(t-\bar{\tau})I(t)dt, \tag{5.5}$$

here $I(t)$ is the registered signal and t_0 the measurement time.

For signals with Gaussian statistics (Gaussian field statistics arises as a result of the central limit theorem of probability theory if the field can be represented as an ensemble of statistically independent random variables [411, 412]), between the correlation functions of the first and second order, there exists the so-called Ziegert relation, Eq. (5.6) [412]

$$g^{(2)}(\bar{\tau}) = 1 + \left| g^{(1)}(\bar{\tau}) \right|^2, \tag{5.6}$$

which is important due to the fact that it makes it possible to calculate the spectral characteristics of first order according to correlation functions of second order or intensity correlation functions (for correlation spectroscopy of optical fields with non-Gaussian statistics, see e.g., [413–416]). In general, the intensity correlation function $g^{(2)}(\bar{\tau})$ can be written as [412]

$$g^{(2)}(\bar{\tau}) = 1 + \left(\frac{\langle I_1 \rangle}{\langle I \rangle}\right)^2 \left(g_s^{(2)}(\bar{\tau}) - 1\right) + 2\left(\frac{\langle I_1 \rangle}{\langle I \rangle}\right) g_s^{(1)}(\bar{\tau})\cos(\bar{\tau}\Delta\omega), \qquad (5.7)$$

where

$$g_s^{(1)}(\bar{\tau}) = \langle f(0)f(\bar{\tau})\rangle, \quad g_s^{(2)}(\bar{\tau}) = \langle f^2(0)f^2(\bar{\tau})\rangle \qquad (5.8)$$

are correlation functions of first and second order, correspondingly, describing the scattering field. It is assumed that $\langle f^2 \rangle$ is normalized to unity.

A nonzero frequency difference, $\Delta\omega$, can arise due to the motion of the scattering medium as a whole, leading to a Doppler shift, and as a result of a specially created frequency shift of the reference beam. If the frequency shift is large enough the power spectrum can be separated into two components, corresponding to the terms $g_s^{(2)}(\bar{\tau})$ and $g_s^{(1)}(\bar{\tau})$ in Eq. (5.7). The first (homodyne) component determined by the fluctuation intensity spectrum of the studied signal, centered on the zero frequency, corresponds to the auto-beating signal of scattering and, therefore naturally exists even in the absence of the reference signal. The second (heterodyne) is centered on the frequency $\Delta\omega$ and is proportional to the spectrum of the intensity of the scattered field. This component can be separated by either artificially creating a frequency shift or by increasing the intensity of the reference beam (see, e.g., [417]). The advantage of heterodyning is that, firstly, this method allows one to obtain information about the optical spectrum, independently of the studied field statistics, and secondly it has a higher signal-to-noise ratio and therefore is more effective in the case of weak signals [246, 247, 271, 418, 419]. However, when applied there always exists the danger of appearance, not of complete, but of partial heterodyning. Detailed investigation of this effect was especially carried out in [420, 421]. The authors of [421] showed that even in this case it is possible to get correct results, but nevertheless, it is usually preferred to avoid the complications connected with this method. In any case, unless extremely necessary (e.g., in the case of non-Gaussian statistics) there is no sense in applying this variety of dynamic light scattering in view of its significant associated experimental difficulties. A more detailed and complete illumination of these problems can be found in [391].

Having in mind what was previously said, the homodyning method was used in all experiments discussed below. Here we took into consideration the fact that Gaussian statistics of scattered light, which is necessary for its application, also determines multiple scattering [369, 379] (see below for more details).

5.2.1.5 Cumulants of the Correlation Function

When the scattering multiplicity is increased, the time correlation function (just like the frequency spectrum) becomes more complicated. Thus, if for single light scattering on thermal fluctuations or monodispersed Brownian particles it appears as a single exponent and as a straight line in the semilogarithmic scale, then the presence of even partial double scattering leads to a significant deviation of such a graph from linearity (see, e.g., [350]). This is connected to the fact that with double scattering, light in the studied beam can fall in different ways, undergoing scattering at different angles (and not at a strictly defined angle as it takes place at single scattering). Scattering at different angles leads to a diverse spectrum broadening, and correspondingly to a differing decay index of the correlation function, because the wave-scattering vector depends on the angle (see Eq. (4.3)). Therefore in this case, exponents with different decay indices contribute in different ways to the correlation function of the analyzed beam. The effect of this on the correlation function is similar to the effect of the polydispersity of the particle size distribution and it arises when it has a nonzero second cumulant. The transition to even larger optical thicknesses and, correspondingly, to even larger scattering multiplicities should be, as it seems, accompanied by further growth of the second and following cumulants. Therefore, in going over to optical thicknesses much larger than unity, one would expect a variety of changes of the type of correlation function right up to the complete disappearance of its time dependence, which would mean the absence of the correlation of photons [382–385]. However, the results of research [386, 388, 390, 391, 393, 394] show that in all cases the time correlation function of multiple scattered light, represented in a semilogarithmical scale (see Fig. 5.2), can be approximated by a square parabola within the accuracy of the experiment, i.e., it allows one to separate, at least, the first and second cumulants (see below for more details).

The measurable correlation function can, in principle, be represented as composed of numerous elementary exponents $\exp(-\Gamma\bar{\tau})$ with values of the parameters Γ distributed from zero to infinity. Then, the first cumulant is equal to this distribution average spectrum half-width; the second to dispersion, and the value of the third cumulant is the measure of asymmetry of this distribution. As experience shows, by increasing the accumulation time one can sometimes achieve a third cumulant significantly distinct from zero and thereby increase the accuracy of the

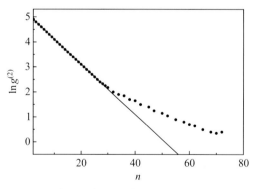

Figure 5.2 Shape of a typical correlation function of multiple light scattering; *n* is the number of the channel of the correlator [391].

data. However, in case of multiple scattering, as it is not possible to give it some defined physical meaning, there is no point in even trying to achieve an accuracy sufficient to determine the cumulant.

The ratio of the second cumulant (the parameter C in Eq. (5.9)) to the square of the first one $(\Gamma_m)^2$ does not usually exceed 0.2. Despite the wide variation in all the experimental conditions this ratio really does not change within the limits of the experimental precision of 20–30%. Thus, the experiment demonstrates that under conditions of unlimited growth of the scattering multiplicity the shape of the correlation function, ultimately, not only does not change essentially, but, on the contrary, achieves a certain stability. Thus, it becomes clear that the first cumulant contains the largest information. Its value can be determined with a margin of error of a few percent, while for higher-order cumulants the inaccuracy is significantly higher. Under conditions of high scattering multiplicity the first cumulant is extremely sensitive to changes in size and concentration of the scattering particles, and also to the observation angle and the size of the cuvette [386, 388, 390, 391, 393, 394].

For the sake of simplicity, instead of the long term "the first cumulant of the time correlation function" we shall use the term "average spectrum half-width" or, more often, simply "spectrum half-width." Taking into account the two cumulants obtained in the experiment the correlation function can be approximated by the following expression:

$$g^{(2)}(\bar{\tau}) = A \exp\left(-2\Gamma_m\bar{\tau} + 2C\bar{\tau}^2\right) + \text{const.,} \tag{5.9}$$

here A is a constant, Γ_m and C are the required parameters. This expression is similar to the one which is traditionally used in the cumulant method [422] to describe single scattering in polydispersed particle ensembles when the correlation function is represented as the sum of a large number of exponential functions with different values of the exponents. In the exponents, commonly no more than two terms are usually introduced, but from time to time also subsequent terms of the expansion in powers of $\bar{\tau}$ are held back. In multiple scattering even in monodispersed particle ensembles there are also numerous exponents with various values (see also Eq. (5.41)). In this case, however, it is not due to the difference of the diffusion coefficients for different sizes, but to the spread in scattering vector values caused by the nonidentity of its separate acts. Nevertheless, these situations are mathematically equivalent. Therefore, the cumulant method, which is well established for analyzing polydispersity (concerning the further development of the cumulant method see, e.g., [416, 423] and references therein), can be used when analyzing the correlation function of multiple scattering [391, 393, 394] just as successfully. In accordance with this method, the quantity Γ_m is the basic parameter determined in the approximation process, and it has the meaning of a mean half-width spectrum of multiple scattering. It is determined by the slope in the graph $\ln g^{(2)}(\bar{\tau})$ vs $\bar{\tau}$ for small values of $\bar{\tau}$.

The shape of the correlation function can appear quite complex at multiple scattering since it depends not only on the scatterer's properties but also on the geometry of the scattering medium. As the modern theory of multiple scattering does not allow us to determine the exact form of the correlation function, we have to restrict ourselves to a formula of the type of Eq. (5.9). It is understandable that for an adequate approximation two members of the expansion in the exponent values may not be sufficient. In addition there are possible apparatus distortions of the correlation function with small $\bar{\tau}$ due to afterpulses of the PM (see below for more). Therefore, it is necessary to specially control the adequacy of the description of the results at each measurement. This control is carried out as a graph of the deviation of the experimental points from the approximated curve. This graph, as has already been mentioned, can be observed on the display screen. With an adequate description there should not be any noticeable systematic deviations in the random background. If the description is found to be inadequate the errors can be eliminated, depending on the reason for its appearance, using the corresponding correction: the introduction of a subsequent term into the approximation, excluding, some initial channels from consideration, etc. As it turned out, in the overwhelming majority of cases with multiple scattering it is more than enough to consider just the linear and quadratic terms of the series. The statistical deviation of the determination of the value of Γ_m did not exceed a few percent and the systematic deviation was even lower [391, 393, 394].

5.2.1.6 Afterpulses

Visual control of the deviation curves makes it possible to detect considerable distortions of the correlation function which can arise in the correlator's first channels when the correlation function argument is several microseconds. Such distortions have been described in literature (see, e.g., [424]) and are connected to the presence of PM afterpulses. To illustrate this phenomenon, approximation results are shown in Fig. 5.3, in the cases where the influence of afterpulses appears more significant.

Afterpulses arise as a rule due to positive ions which the electronic avalanche, caused by photons on the photocathode, in some cases dislodges from PM dynodes surfaces. Moving in the opposite direction to the photocathode they also dislodge electrons. These electrons, no longer being caused by photons of scattered light, create false signals on the PM's output, the secondary impulses. The delay of such an afterpulse is determined by the time of ion transit, which in turn depends on the power supply voltage of the PM and its geometric dimensions. Therefore, it is not possible to substantially change the time delay of the afterpulses (for PM 79 this time is only a few microseconds). As not every PM impulse is accompanied by an afterpulse (in the overall flow they represent only a fraction of a percent) the distortion of the correlation function due to afterpulses is only significant under certain conditions.

With an increase in the scattering multiplicity, the spectrum broadens substantially and, correspondingly, the correlation time of the intensity of scattered light (size, inverse spectrum half-width) decreases. When the correlation time becomes several times less than the average time between photocounts, the ratio of photocounts correlating with each other decreases sharply. Only a small part of the photocounts is divided into time intervals less than or almost equal

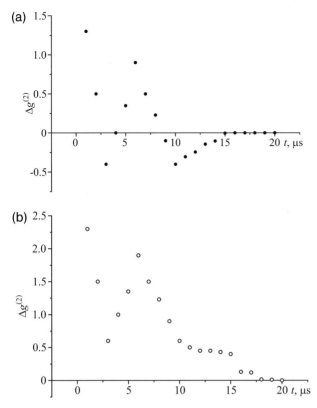

Figure 5.3 (a) Deviation of the correlation function from its approximation according to data from all channels (top). (b) The correlation function of light scattered on stationary scatterers (bottom).

to the correlation time and it contributes to the time-dependent part of the correlation function. The rest, the main part of the impulses-photocounts contributes to the correlation function "pedestal" which does not depend on time and so does not contain useful information. Against the background of rare correlations of principle impulses, their correlations, together with afterpulses, become substantial.

An increase in the intensity of scattered radiation usually leads to a reduction in the negative influence of afterpulses. This is because the average time between impulses-photocounts decreases and becomes small again compared to the correlation time. However with multiple scattering, it turned out that on the parts of the scattering medium surface, where intensity is minimal, the spectrum half-width reaches the highest value. This is easy to understand as the few photons which reach these parts undergo the largest number of scattering acts and a higher mean scattering multiplicity leads to a bigger spectrum half-width.

If we exclude from the approximation the data which were gathered in the first channels, where the largest distortions can be seen on the deviation curve (see Fig. 5.3(a)), then such a corrected correlation function can nevertheless be used. Unfortunately, the correlation function for small values of its argument, which had to be discarded due to such a correction procedure, contains greater valuable information about its first cumulant. However, if there are afterpulses the loss of this part of information is unavoidable when using a single PM. However, there exists a method which allows us to exclude the influence of afterpulses on the registered correlation function, the cross-correlation method, which uses not one but two PMs simultaneously (see Chapter 6 for more details).

The validity of the described procedure can be proved by comparing the deviation graph, arrived at through corrections, with the light-intensity correlation function scattered on a fixed opaque screen. The part of such a correlation function which depends on time should be subject to only PM afterpulsing as photons scattered on a fixed screen do not correlate with themselves. Comparing Fig. 5.3(b), which shows the result of such measurements, with Fig. 5.3(a), confirms the validity of the described procedure of data correction.

5.3
Physical Modeling

5.3.1
Model Systems

Experimental study of the principles inherent in multiple light scattering is natural to be performed on the same model systems which are widely used to study single scattering. In the latter case, systems of water suspensions of submicroscopic polystyrene spherical particles (latexes) are mostly employed. The special technology used for producing polystyrene latexes guarantees a high degree of monodispersity (deviations in the size of the scattering of particles take up only a few percent). Polystyrene particles have a high degree of sphericity, which is testified by electron microscopy [425, 426]. In addition, the density of polystyrene differs from the density of water by only 5%, and therefore the rates of changes in the concentration of the particles, due to its sedimentation, are quite low. Thus, in many experiments this effect can be simply ignored. The refraction index of polystyrene relative to water is quite high (equal to 1.2), which guarantees an extremely high scattering intensity in the latex.

Latex production technology (small amounts of emulsifier are added for stabilisation) guarantees the conditions which eliminate the process of particle coalescence. Therefore, even with a concentration of a few volume percent it can be assumed that the suspended matter consists of separate Brownian particles, whose effective size does not depend on concentration. Due to these peculiarities polystyrene latexes are traditionally used as model systems for studying single and double light scattering (see, e.g., [350, 421, 425, 427, 428]). As light absorption in polystyrene is small, scattering multiplicity in latexes can be very high (>1000). All these properties allow us to consider latexes as great model systems for studying the patterns of not only single and double scattering, but also of high-multiplicity scattering, including the one close to the critical point.

As it turned out [383, 386], an essential difference between multiple scattering and single scattering is the dependence of the correlation function on the concentration of the scatterer. Particle concentration in the model system is determined in two ways [386, 390], both of which give similar results. The first method is to weigh the latex before and after evaporation, which, as their radius is known, allows us to determine the volume concentration of the particles. The second method is connected to the measurement of the extinction coefficient in a thin latex layer and the subsequent calculation of concentration. To make such a calculation the scattering efficiency factor has to be determined using the well-known light scattering tables [429, 430]. When the concentration of the latexes changes within the limits 10^{-3} to 10^{-1} of the volume of polystyrene, the relationship of the cylindrical cuvette's diameter to the extinction length changed from 10 to 10^3.

5.3.2
Dependence of the Spectrum Half-width of Multiple Scattering on the Physical Characteristics of the System and on the Scattering Multiplicity

5.3.2.1 Dependence on the Viscosity of the Fluid
The mobility and diffusion rate of Brownian particles depends substantially on the viscosity of the medium in which the particle is moving and, consequently, on its temperature. The spectrum half-width of single light scattering on Brownian particles Γ is, as a consequence of Eq. (4.18), proportional to the coefficient of particle diffusion and, therefore, by virtue of the Stokes–Einstein formula, also depends on temperature and medium viscosity as

$$\Gamma = Dq^2 = \frac{k_B T}{6\pi\bar{\eta} r_h} q^2 = \Gamma\left(90^\circ\right)\left(1 - \cos\theta\right), \tag{5.10}$$

where D and r_h are the diffusion coefficient and hydrodynamic radius of the Brownian particles, respectively. The hydrodynamic radius differs from the particle radius by so-called coat, captured in the course of the motion, which are made up of a layer of molecules of a surface active agent, dispersed media, etc. (see, e.g., [425]). The other symbols have the same meaning as in Eqs. (4.2), (4.3), and (4.29).

Equation (5.10) shows that if the radius of the particles carrying out a Brownian motion in a certain medium is known, then one can get the value of the viscosity coefficient of this medium from the experimental definition of the spectrum half-width. Therefore, measurements of the spectrum half-width of single scattering in especially introduced scatterers are often employed to determine the medium viscosity in a nondestructive way [418, 419, 431, 432]. For the first time, such researches in conditions of high multiplicity scattering were conducted in [390, 391] (see also [393, 394]). In these experiments, polystyrene Brownian spheres with a radius of 280 nm were placed into a cylindrical cuvette with a diameter of 20 mm and height 40 mm. The concentration of the particles was such that the extinction length, determined by the measurement of the transmission of light

Table 5.1 Parameters Γ_m and the ratio (Γ_m/Γ) in dependence on temperature.

t (°C)	Γ_m (kHz)	(Γ_m/Γ)
35	10.9	28.4
45	13.9	29.1
55	17.0	29.1
65	19.7	28.1
75	22.5	27.3
85	27.7	28.8

in a thin layer of latex, was 0.27 mm. This value, which is much smaller than the cuvette's diameter, fully guaranteed high multiplicity scattering. At the same time, the average distance between the particles, estimated according to the value of the volume concentration of polystyrene (1.6×10^{-3}), was 15 times larger than the radius of the particles. This allowed us to consider the latex suspension as an ensemble of independent Brownian particles. The cuvette with the latex was placed into a thermostat. The deviation in the temperature from the average was not larger than 0.05 K. The time correlation functions of multiple light scattering were measured at different latex temperatures with an observation angle of 90° using a correlation spectrometer.

The values, obtained for the spectrum half-width, Γ_m, are shown in Table 5.1. In the last column, the value of the ratio (Γ_m/Γ) is given in dependence on temperature. It is evident that this ratio does not depend on temperature as the observed dispersion of a few percent does not have a systematic character and fully fits into the errors of measurement of Γ_m. This, in turn, means that the temperature changes of Γ_m, when light is scattered on Brownian particles, also, just as for single scattering, are only determined by temperature changes of fluid viscosity when there are suspended particles in it [391, 393, 394]. Thus, it is clear that the temperature dependence of viscosity on the multiple light scattering spectra on Brownian particles can be judged exactly in the same way as it was done in the case of single scattering in [432]. In Fig. 5.4 the temperature dependence of water viscosity is shown, determined by using the data given in Table 5.1. The normalisation was carried out using the value of viscosity for 85 °C.

Changes in the temperature of the liquid did not lead to any specific differences in the behavior of the spectrum half-width between single and multiple scattering systems. The only difference consists in the fact that the values of Γ_m are by several orders of magnitude higher than the values of the spectrum half-width of single scattering, because of high scattering multiplicity, other factors being equal. The results obtained agree with the idea first formulated in [383], according to which Γ_m can be represented as the product of the average scattering multiplicity by $\Gamma(90°)$. As a result of the investigations carried out in [390, 391, 393, 394] an important conclusion was made from a theoretical as well as practical point of

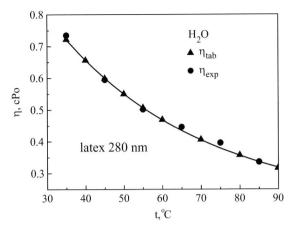

Figure 5.4 Temperature dependence of water viscosity using data from correlation spectroscopy of multiple scattering (•) compared to the date given in Table 5.1 (▲).

view: the spectrum half-width of multiple scattering on Brownian particles (Γ_m), as in the case of single scattering, is proportional to the particle diffusion coefficient, i.e., $\Gamma_m \sim (T/\overline{\eta})$. However, the proportionality coefficient cannot be calculated as easily as for single scattering in this case.

In conclusion, the following should be mentioned. Traditional methods of measuring transport properties are mainly connected with the introduction of gradients of the respective quantities into the investigated system. However, the various perturbations which arise in the medium are not easily adequately accounted for. At the same time, light scattering methods, as applied to the considered problem, are free of such inadequacies as they do not suppose the presence of any macroscopic gradients. With respect to this peculiarity and also due to the increase in the availability of the correlation spectroscopy method, dynamic scattering methods have recently become more and more widespread. In relatively recent papers [418, 433] the viscosity of water was also measured using special scatterers. Unfortunately, the measurements were only performed at room temperature [418] under conditions of single scattering and they did not coincide very well with the literature data.

It is worth remembering that between 1966 and 1970, E. Matizen and colleagues in their pioneering works used microphotography to observe Brownian particles close to the critical point in binary mixtures [431, 434]. Then, Lyons et al. [435] used correlation spectroscopy to determine the anomalies of viscosity in the weakly opalescing mixture nitroethane-isooctane close to the critical point using a teflon sphere placed in the mixture. In 1980, Anisimov et al., in the previously cited study [432], conducted a very serious and useful methodological investigation on the influence of various factors such as multiple scattering, non-Gaussian statistics, and the polydispersity of Brownian particles on the precision of determining viscosity using the homodyning method. They also discovered the gradual decrease in the angular dependence of the spectrum half-width when scattering multiplicity increases, until it completely disappears when the latex concentration is 6×10^{-4}, that is "with complete multiple scattering." The authors drew this conclusion from the visual picture. However, as was subsequently shown [386, 390, 391, 393, 394], the angular dependence disappearance actually occurs because such a concentration, being intermediate, really does not guarantee a sufficient scattering multiplicity (see, e.g., Figs. 5.7 and 5.8). As for the results of [386, 390, 391, 393, 394], which were obtained via multiple scattering, they allow us to widen the range where dynamic scattering for measuring viscosity using the nondestructive method can be applied, including the large class of concentrated systems with high extinction and where the behavior of the viscosity can also be non-Newtonian.

5.3.2.2 Dependence on the Optical Thickness of the Scattering Medium

In multiple scattering, it is the medium optical thickness that plays a significant role, as it determines the scattering multiplicity. The optical thickness depends on geometric dimensions of the disperse medium and the concentration of suspended particles, as much as on the effectiveness of light scattering on them [429, 430]. This effectiveness, in turn, depends on the ratio of particle size and wavelength and on the particle-to-medium refraction index difference. Therefore, it is significantly easier to study experimentally the dependence of the spectrums half-width on the optical thickness of the disperse medium by varying the particle concentration and the geometric factor, separately. The following part is devoted to the results of our study carried out in such a manner [391, 393, 394, 436–438].

5.3.2.3 Angular Dependence

One of the simplest well-known properties of the spectrum half-width of single scattering, Γ, is its proportionality to the square of the wave vector of scattering [319]. According to Eqs. (4.3) and (5.10), the angular dependence of Γ can be represented as

$$\Gamma \sim \sin^2\left(\frac{\theta}{2}\right). \tag{5.11}$$

There are a lot of well-known papers (see, e.g., [383, 384, 432]) where with transition to multiple scattering, a leveling out of the angular dependence with a tendency toward its gradual disappearance was observed experimentally (this fact was established in these studies exclusively visually, see also Section 5.4). Therefore, the literature at that time confirmed the opinion that with high multiplicity scattering the angular dependence of the spectrum half-width should disappear completely (see, e.g., the review [350]).

In general, the absence of an angular dependence at high multiplicity scattering is rather more surprising than natural. It is true that all experiments indicate that the spectrum half-width depends on the scattering multiplicity in one way or another. But the scattering multiplicity in a cylindrical cuvette cannot be the same for different observation angles, even if for no other reason than the fact that the effective optical thickness of the medium depends on the angle. Thus, it seems to be clear that the larger the optical thickness, the higher the scattering multiplicity and, consequently, the greater the spectrum broadening. As broadening already occurs in the first scattering act, it is not clear why the subsequent acts should not lead to a further increase in the spectrum half-width. Concerning the irrefutable experimental fact that with a not very high scattering multiplicity the angular dependence is in fact absent, we can only suggest that with a well-defined combination of cuvette radius, concentration, and optical characteristics of the particles, it is clearly very likely that the photon leaves the scattering medium through any surface point in approximately the same number of "steps."

The experiments [386, 390, 391, 393, 394, 436–438] confirmed the validity of all these considerations. Let us consider the angular dependence of the spectrum half-width in latex with various sizes and concentrations of particles. A cylindrical glass cuvette (with diameter 20 mm) was filled with latex to a height equal to twice its diameter and so the focussed laser beam was incident in the middle of the latex. The diameter of the beam did not exceed 0.5 mm. The analyzed scattered light beam was selected in the horizontal plane and coming out along the radius of the cuvette at the same height at different angles (observation angles) in relation to the incoming beam. The diameter of the part of the surface of the scattering medium where the light was incident on the photon detector was 0.3 mm, i.e., compared to linear dimensions of the medium it was small.

The measurement results of the spectrum half-width of multiple scattering (Γ_m) in concentrated latex with particles with a radius of 280 nm are shown in Fig. 5.5. For comparison the full line represents the angular dependence of the spectrum half-width of single scattering (Γ), which was observed in the same, but highly diluted, latex. The quantity $\sin^2(\varphi/2)$ is given on the horizontal axis, as it is usual for single scattering. It is clear from the graph that Γ_m monotonically decreases with an increase in the observation angle while at the same time Γ increases. As we can

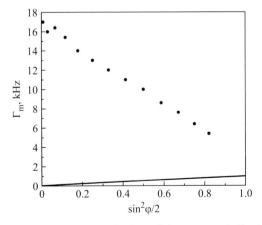

Figure 5.5 Experimental values of the spectrum half-width of multiple scattering (Γ_m) in concentrated latex with a particle radius of 280 nm (•). The full line is the angular dependence of the spectrum half-width of single scattering (Γ) in the same latex, however, highly diluted [391, 438].

see, this result did not correspond to the opinion dominating at that time that the spectrum half-width of multiple scattering should not depend on the observation angle (see, e.g., [350, 383, 384]).

5.3.2.4 Dependence on the Polarization Mode

As we can see from Fig. 5.5, the spectrum half-width of multiple scattering can exceed the spectrum half-width of single scattering by an order or more. Moreover, this assertion relates to the half-width, measured with crossed and also with parallel positioned polarizers. It is known that in the case of single scattering the spectrum of polarized scattered light contains information about the translational motion and the unpolarized spectrum contains the same about rotating motion [336]. The half-width of such spectra can be substantially different. In the case of high multiplicity scattering, the spectrum half-width did not appear much greater with crossed polarizers than with parallel ones. With a decrease in the observation angle, both half-widths grow and the difference between them (with a relative measurement error of 3–5%) becomes almost invisible.

To illustrate this result, we can use the measurement data shown in Fig. 5.6 for latex with a particle radius of 80 nm and volume concentration 3.18×10^{-3} at different observation angles. It is clear that the growth of the spectrum half-width when the observation angle is changed is connected to the growth in the scattering multiplicity, which is accompanied by a decrease in the relative difference between the spectrum half-width of polarized and depolarized scattered light. Therefore, we should mention another peculiarity of multiple scattering: when multiplicity is increased the ratio between the spectrum half-widths of polarized and depolarized scattered light approaches unity. Most of the measurements in [386–395, 397, 398] were conducted using only cross position of polarizers. This is another way of excluding partial heterodyning [421], which could arise during the measuring due to accidental parasitic highlighting of the cuvette's wall. It should also be noted that

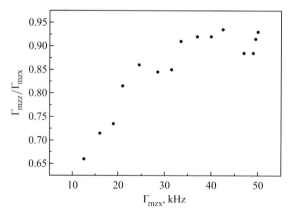

Figure 5.6 Comparison of the half-width of the polarized and nonpolarized components of the scattering spectrum [391].

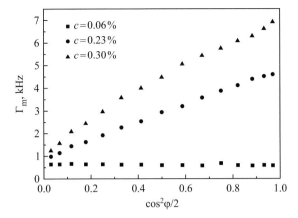

Figure 5.7 The angular dependence of the spectrum half-width of multiple scattering for latex with a particle radius of 770 nm [386].

it is the depolarized component which is formed by high multiplicity scattering. It appears that this is where their differences lie under high multiplicity scattering. Thus, unless otherwise specified, the spectrum half-width of multiple scattering will be assumed to have a half-width value measured using cross polarizers.

To find out how the results of angular measurements of Γ_m (Fig. 5.5) correlate with the results of [385], in which the absence of an angular dependence of the depolarized component of the multiple scattering spectrum half-width was experimentally demonstrated, measurements of the angular dependence of Γ_m at different latex concentrations were carried out [386]. Three latex samples with a particle radius of 770 nm and concentrations of 3.0×10^{-3}, 2.3×10^{-3}, and 0.6×10^{-3} were used. The corresponding extinction lengths were 0.21, 0.24, and 1.93 mm. The results of measurement of Γ_m are shown in Fig. 5.7. It is clear that when the concentration increases from very small values (single scattering) the angular dependence of the spectrum half-width disappears at first, but then reappears, and moreover, the sign of its slope becomes the opposite one. This makes it clear why the angular dependence of the kind shown in Fig. 5.5 is not observed during the experiment [385]: the maximum scatterer concentration used did not exceed 0.5×10^{-3} and consequently, the extinction coefficients were small. It should be noted that the value of the maximum concentration of the latex coincided in all three studies [382, 384, 432]. For a 10-mm side square cuvette [385], the coefficients corresponded just right to such an intermediate case of not very high scattering multiplicity (about three).

At the same time, the data corresponding to the concentrations 3.0×10^{-3} and 2.3×10^{-3} in Fig. 5.7 show that the slope of the angular dependence of Γ_m depends substantially on the concentration of the scatterer. However, the shape of the dependence does not undergo any significant changes and stays close to a straight line. We will look at this in more detail in the next part, where we will discuss the concentration dependence of Γ_m. We should also mention that in the case of single scattering the only characteristic depending on the concentration of the scatterer is the intensity of scattered light. The spectral characteristics of single scattering do not depend on concentration. It should also be emphasized that we are talking about independent scatterers whose motion does not depend on concentration. The effects connected with the slowing down of the Brownian motion, due to their interaction, for large particle concentrations are not covered here.

For low multiplicity scattering, when it is close to double, a dependence of the spectrum half-width on the concentration of the scatterer was observed by the authors of [383, 384]. They found that the spectrum half-width is proportional to the concentration. Without, in effect, conducting experiments with high multiplicity scattering (although considering them as such) and lacking any valid argument, they proposed that such proportionality would remain true even with an increase in the scattering multiplicity. The true situation of multiple scattering will be discussed in the next section.

5.3.2.5 Dependence on the Concentration of the Scatterer

In the previous section, we discussed the character of the angular dependence of Γ_m upon transition to high multiplicity scattering. The concentration dependence of Γ_m could also, in principle, have undergone changes. To check this, a special experiment was conducted [437] (see also [394]).

When studying scattered light in concentrated latexes, it should be kept in mind that for high concentrations the interaction of particles can become noticeable. Such latexes cannot be regarded as a suspension of independent Brownian particles and, consequently, cannot be used as a model system. Therefore, when studying the concentration dependence of Γ_m the possibility of varying the concentration of the scatterer is very restricted. The upper limit suitable for concentration experiments is determined by changes in the particle diffusion coefficient, whose size becomes bigger than the measurement error of the spectrum half-width. Such a concentration can be estimated using the dependence of the coefficient of diffusion, D, of an ensemble of hard spheres obtained in [439]

$$\frac{D}{D_0} = 1 - 1.73c + 0(c^2), \tag{5.12}$$

where c is the concentration of the particles in volume units. Using Eq. (5.9), we find that the concentration $c = 1.7 \times 10^{-2}$ corresponds to a decrease in the diffusion coefficient by 3%. Therefore, with a margin of error in the measurement of Γ_m of just a few percent, it was necessary to use latexes with a concentration not exceeding 10^{-2}.

The suitable lower limit for such concentration experiments is determined by the need to obtain a value for the extinction coefficient which guarantees sufficiently high scattering multiplicities, for which one can observe the angular dependence of Γ_m. As the extinction coefficient depends substantially on the dimensions of the scatterer and the scattering multiplicity depends on the dimensions and form of the scattering medium it is impossible to uniquely determine the lower limit of acceptable concentration. As an example, we can mention that, according to the data shown in Fig. 5.7, even for latex with a particle radius of 770 nm with a concentration of 0.6×10^{-3} an angular dependence of Γ_m is absent. Consequently, the scattering multiplicity is low despite the fact that polystyrene particles of this size scatter light in water quite effectively. Therefore, the range of suitable concentrations was quite narrow and did not exceed one order of magnitude (from 10^{-2} to 10^{-3}). So it was important that the ratio of the concentration of the latex samples could be determined with great precision (in the experiment the error did not exceed a tenth of one percent). This was achieved by controlling the water dilution of the original latex, whose initial concentration was determined by evaporation (with only a small decrease in precision).

Measurements of the spectrum half-width of multiple scattering (Γ_m) were conducted on latex samples with a particle radius of 80 nm ($\Gamma(90°) = 0.92$ kHz), placed into a 20-mm diameter cylindrical cuvette, as was already mentioned. The measurements shown in Fig. 5.8 were carried out at different observation angles for five particle concentrations: $c_1 = 4.39 \times 10^{-3}$, $c_2 = 3.18 \times 10^{-3}$, $c_3 = 2.69 \times 10^{-3}$, $c_4 = 2.17 \times 10^{-3}$, $c_5 = 1.13 \times 10^{-3}$. These concentrations correspond to the following extinction length values: 0.61, 0.85, 1.0, 1.2, and 2.4 mm. As can be seen in the figure, the angular dependence of Γ_m, typical for high scattering multiplicity, appeared only in four of the more concentrated samples. These data were used for further analysis using the special graph shown in Fig. 5.9. The points on the graph which lie along one vertical correspond to the same value of observation angle. By showing them in this manner, we get rid of the explicitly unknown, but always the same in conditions of geometric similarity, coefficient and draw out the dependence of Γ_m on concentration in an explicit form. It should be mentioned that when the horizontal coordinates increase the

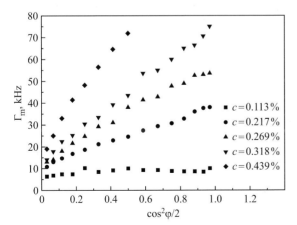

Figure 5.8 Angular dependence of the half-width, Γ_m, of the spectrum of multiple scattering for a latex with a radius of the particles equal to 80 nm [391–394, 437].

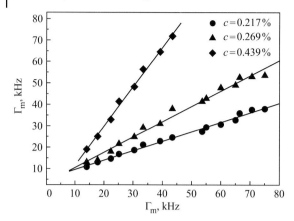

Figure 5.9 Same dependence as in Fig. 5.8 but the values of Γ_m for one of the concentrations are chosen as an argument. Their horizontal coordinates equal the corresponding values of Γ_m for a concentration c_2 and the vertical coordinates equal to Γ_m for three other concentrations [391–394, 437].

observation angle steadily decreases and in the extreme points the angles are $160°$ and $20°$, respectively.

The experimental data in Fig. 5.9 form three straight lines, crossing at the same point, whose coordinates are both identical. This result allows us to conclude that the angular dependencies of Γ_m obtained can be considered as similar provided that some constant term ($\Gamma_0 = 8$) is accounted for, which does not depend on the scatterer's concentration. In our case, it equals $\Gamma_0 = 8$ kHz.

The values of the slopes of the straight lines in this figure contain information on the concentration dependence of Γ_m. The calculation by least-squares method results in the following values: 0.45, 0.77, and 1.8, which well coincide within the experimental error (5–10%) with the squares of the concentration ratios equal to 0.47, 0.71, and 1.9. To demonstrate this fact, the straight lines in Fig. 5.9 are traced with a slope precisely equal to the squares of concentration ratios. It is clear that the experimental points agree with these lines. Thus, the quadratic dependence of Γ_m on the concentration of the scatterer was discovered under high multiplicity scattering. This result is shown in more detail in Table 5.2, where $b_{i,j}$ ($i, j = 1, 2, 3, 4$) are the values determined from experimental data for the concentrations c_i and c_j and represent the coefficients of similarity transformation of one angular dependence to another. To calculate these values the results of more than 50 measurements of Γ_m were used altogether.

The results shown in Table 5.2 convincingly demonstrate the quadratic concentration dependence of Γ_m (taking into account the contribution of Γ_0 which depends neither on concentration nor on observation angle). As can be seen, the greatest deviation between the $b_{i,j}$-values and the squares of concentration ratios does not exceed $\approx 10\%$ (in most cases it is significantly lower) and can be completely accounted for by the error in the measurement of Γ_m ($\approx 5\%$), as the

Table 5.2 Values of the coefficients, $b_{i,j}$, for different concentrations, c_i and c_j.

i,j	1,2	1,3	1,4	2,3	2,4	3,4
$\frac{c_i^2}{c_j^2}$ (calc.)	1.90	2.67	4.08	1.40	2.15	1.53
$b_{i,j}$ (exp.)	1.8	2.4	3.9	1.3	2.2	1.6

errors in determining particle concentrations and observation angles were much less. This leads us to the inevitable conclusion that the linear concentration dependence of Γ_m, which was revealed in previous papers [383, 384], accounts for only slow scattering multiplicity but changes to quadratic one at the transition to high concentrations and, therefore, to high scattering multiplicity.

These results allow us to make some further conclusions about the angular dependence of Γ_m. In particular, the fact that the experimental points on Fig. 5.9 lay on straight lines, indicates the presence of a universal type of angular dependence of Γ_m in the area of both high particle concentration and high scattering multiplicity. If the concentration dependence is known, then it is not difficult to place all the measured values on one curve using the following transformation

$$\Gamma_{mi} = \frac{c_i^2}{c_j^2} \left(\Gamma_{mj} - \Gamma_0 \right) + \Gamma_0, \tag{5.13}$$

where Γ_{mi} is the spectrum half-width for the concentration c_i, measured using some values of the observation angle and Γ_{mj} is the spectrum half-width measured using the same observation angle but for the concentration c_j.

Figure 5.10 shows the results of such a transformation presented in the manner typical for the concentration c_2. Despite the fact that among the experimental points, corresponding to the same observation angle, there is some dispersion, one can nevertheless draw a smooth curve which, for virtually the whole of its length,

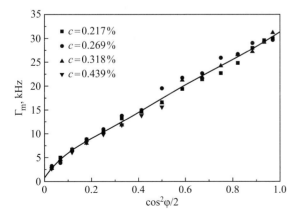

Figure 5.10 Generalized angular dependence for different concentrations.

does not differ significantly from a straight line. An exception is found, quite reasonably, in the range of large observation angles, close to the "back" scattering direction where the line has a noticeable curvature. It is also typical that here such a dispersion of the data points becomes significantly smaller, which once more shows that the value of Γ_0, to which the values of Γ_m tend in this range, does not depend on concentration (for more details see below).

At the beginning of this section it was mentioned that the effect of concentration on Γ_m reflects the dependence of this quantity on the optical thickness of the medium. We can, therefore, assume that the given results allow us to confirm that Γ_m depends on the optical thickness of the medium quadratically. It should be stressed that the given type of the angular dependence of Γ_m relates to the situation when the laser beam penetrates into a cylindrical cuvette with the scattering medium from a side. When the scattering geometry is different, the type of the angular dependence of Γ_m can also be different. However, as Pavlov [440] showed, we can conclude from analyzing the diffusion equation, taking into account the conditions of its applicability, that the spectrum half-width of multiple scattering in geometrically similar conditions is always proportional to the square of the optical thickness. This result was obtained later while studying multiple scattering using slab geometry [441]. Therefore, the result obtained for cylindrical geometry [386, 390–394, 436–438], i.e., for the most widespread experimental conditions, is quite general. Moreover, in comparison to slab geometry it is very easy to change the mean scattering multiplicity by changing the observation angle in the case of cylindrical geometry. It should also be noted that the important advantage of experiments in slab geometry lies in the fact that it is comparatively easy to investigate analytically (see below and Appendix A.1 for more information).

5.3.2.6 Dependence on the Dimensions of the Scattering Media

The proportionality of the spectrum half-width of multiple scattering to the square of optical thickness was further checked in experiments using cylindrical cuvettes of different diameters, 20, 15, and 13.3 mm. The measurement results of Γ_m for latexes with a particle radius of 280 nm, which were published in [390] and [438], are shown in Figs. 5.11 and 5.12. In Fig. 5.11, the angular dependences of Γ_m are shown and in Fig. 5.12 a presentation is given analogous to that discussed above (Fig. 5.9). The quantity Γ_m is used for the horizontal coordinates, corresponding to measurements in a 15 mm cuvette. The slopes of the straight lines formed by experimental points are equal to 1.78 and 0.64, which agree satisfactorily with the squares of the corresponding diameter ratios (1.78 and 0.79). It is worth mentioning that the crossing of the straight lines in Fig. 5.12 does not happen at the origin of the coordinates and makes it possible to distinguish the contribution to Γ_m which is independent of the cuvette's diameter. This situation is similar to the previously discussed one with Γ_0 contribution (Fig. 5.9). The major difference in the values of these contributions (1.3 kHz and 8 kHz) is connected with the fact that these series of measurements were carried out with particles of different dimensions. Thus, the experiment specifies the existence of the contribution to Γ_m, which

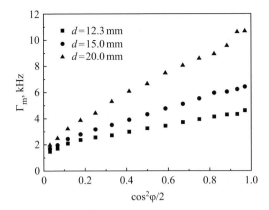

Figure 5.11 Latex with a particle radius of 280 nm and cuvettes of different sizes.

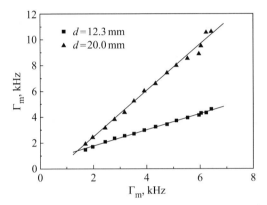

Figure 5.12 Figure similar to Fig. 5.9 for different diameters.

does not depend on the medium optical thickness but on the dimensions of the scatterer.

Figure 5.13 shows the dependence of Γ_m on $\cos^2(\phi/2)$ for latex whose particle radius is determined by the multiple scattering spectrum. According to electron microscopy data this latex had a mean radius of 45 nm, calculated using the results of optical experiments, the hydrodynamic radius was equal to 48 nm (the calibration of the used cylindrical cuvette has been carried out in view of Eq. (5.13) and Fig. 5.10). Due to the "coat" the hydrodynamic radius is always bigger than the "electronic" one and therefore the result can be accepted.

With these considerations, the description of the physical modeling of critical opalescence with monodisperse systems with high and regulated extinction, first conducted in [386, 390–392, 436–438], is completed and we can move on to looking at the theoretical aspects of this problem, the formulation of mathematical models of multiple light scattering on Brownian scatterers. This stage anticipates the formulation of a critical opalescence theory in the close vicinity of a critical point, and Chapter 6 is fully dedicated to this topic.

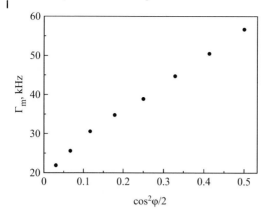

Figure 5.13 Latex with a particle radius of 48 nm [141].

5.4
Mathematical Modeling

All the experimental results obtained by studying light spectra of multiple scattering on model systems of Brownian particles can be explained (with the exception of Γ_0) within the framework of theoretical concepts developed in [141, 386, 390–394, 436–438]. It should be emphasized that even the simplest diffusion model, the model of a random walk of a photon in a high-extinction monodispersed medium, used to interpret results in our first study on this subject [386] was found to be adequate. Further, a more profound approach, the diffusion approximation of radiation-transport theory was used instead of the simple random walk model [354–356]. There is no doubt that this approach added some rigour and improved the quantitative agreement with the experiment. It should be noted, however, that it has not changed the principal conclusions of our first simple model [386]. Therefore, it can be considered that it was in [386] that the foundation was first laid for a new scientific area, correlation spectroscopy of multiple light scattering.

Similar results were obtained a few years later for light scattering in slab geometry in [441]. This method is still intensively developing and has been given the name diffusing-wave spectroscopy [442] (see, e.g., [443, 444, 454–463]). The experimental analyses [441–443] of strongly scattering media were also completed with the help of photocorrelation techniques. In the theoretical papers [445, 446], the diffusion approximation of radiation-transport theory in non-absorbing and weakly absorbing media was also used to study this question. Below, we shall look at the results obtained in these works in more detail (see Appendix A.1).

Now, after this short but necessary introduction, let us move directly on to the essence of the problem and show how the above-discussed results obtained in experiments on model systems (see Section 5.3) can be explained with the help of the diffusion approximation of radiation-transport theory [347–349, 354–356].

5.4.1
The Simplest Diffusion Model Approach

When light passes through a dispersed medium, as a result of quasielastic scattering on chaotically moving Brownian particles the scattered light spectrum widens. With single scattering this phenomenon has been intensively studied and the results are well known (see, e.g., [319, 399–402]). With an increase in multiplicity the scattering picture becomes more complex, and when the medium optical thickness grows so much that the scattering becomes multiple, the situation is not only no longer simple and clear but also requires completely different approaches. The intensity of light passing through randomly distributed nonabsorbing scatterers decreases in accordance with Bouguer's law, a redistribution of energy from incidental primary waves to secondary scattered ones takes place (see, e.g., [355, 464]). There are two approaches to describe this process without violating the law of energy conservation.

5.4.1.1 The First Approach

The electromagnetic field which propagates in a medium is represented by the sum of fields of different scattering multiplicity. It is commonly believed that the electromagnetic field propagates itself in the same way between separate scattering acts as in vacuum. Although the absence of absorption makes this assumption quite natural, its application to turbid media is accompanied with great difficulties. As a matter of fact, the fields of different scattering multiplicities have to interfere with each other. The extinction can only be explained by the interference of an incidental wave with scattering ones, single scattering with double scattering, etc. [464]. It is clear that in the case of multiple scattering, the energy conservation law can be preserved only by taking into account the interference. Unfortunately, this is sometimes overlooked in passing from generally accepted single scattering approximations, where interference is not important, to multiple scattering where it plays a dominating role. As a result, a certain incorrect concept arose [383, 384] which states that chaotic distribution of scatterers leads to the disappearance of the interference between fields of different multiplicity (just as Rayleigh proposed, the noncoherence of the secondary waves, radiated by gas molecules, was caused by the independence of their thermal motion (see Section 4.2.1). This error did not stop Rayleigh obtaining the true result because he was dealing with single scattering in gases).

Applying the energy conservation law to electromagnetic waves scattering on an elastic dipole [465], Max Planck, at the turn of the 20th century, came to the conclusion [466] that in a nonabsorbing medium the refractive index has a small imaginary part in view of the slowing down by radiation. This is what causes the decay of the intensity of the oscillator's vibrations. Introducing the slowing down by radiation in a real system of examination means, in practice, taking into account light extinction depending on scattering.

Modern theories of multiple scattering have further developed Planck's ideas. For example, Ishimaru arrived at the law of extinction of the coherent intensity for particles with an arbitrary scattering indicatrix [355]. In the theory of single scattering the imaginary part of dynamic polarizability is usually ignored. Then, simultaneously with it, as it follows from the theoretical results [355, 466], the interference fields of different scattering multiplicities are not taken into account and therefore, extinction is completely ignored. This can only be done if the optical thickness of the sample is much less than unity, i.e., when single scattering is dominating in the system.

5.4.1.2 The Second Approach

The difficulties mentioned above can be avoided if one takes extinction into consideration, having included it ab initio into the coherent field propagation law. Such an approach [341, 342] has significant advantages since all the effects of interference in a system of uncorrelated scatterers are automatically taken into account and, as a result, a statistical problem in the spirit of the transport theory arises. Within the limits of high multiplicity scattering, this statistical problem can be reduced to the diffusion approximation [347–349, 467]. Thus, the main factor that determines the statistics of high multiplicity scattering of radiation is light extinction. This is the qualitative difference between strongly opalescing systems and those with single scattering [386]. As Ishimaru pointed out [468], it is convenient to describe multiple scattering be means of correlation functions (see below for more details).

Now let us go over directly to looking at a simplified diffusion model of multiple scattering. In the case of independent scatterers, such a model can be based on the following assertions [386, 438]:

- a wave passing through a chain of scatterers leads to the multiplication of time correlation functions which define separate scattering acts;
- the correlation function under study corresponds to the statistical averaging along various chains of scatterers;
- between the different scattering acts attenuated waves propagate with an intensity of $I \sim r^{-2} \exp(-hr)$, where h is the extinction coefficient.

Since such a statistical model corresponds exactly to the transport equation which Ishimaru [355, 468] obtained for the time correlation function, the last statement can be formulated in a different way:

- the statistics of chains is the same as for random walk of classical particles without interference. The interference is accounted for by attenuation.

The multiplication of the correlation functions leads to the additivity of the spectral broadening [383, 384, 386]. As a result, the multiple scattering spectrum half-width can be represented as

$$\Gamma_m = \sum_i \Gamma_i, \tag{5.14}$$

where Γ_i is the spectral broadening at the ith scattering act (this principle is quite easy to justify the simplest situations, while the problem of effectively summing up the fields, running along different chains of the scatterers, is rather difficult to solve).

In Eq. (5.14), each component of Γ_i is determined by the size of the scattering vector in the ith node of the chain. In the extreme case of multiple scattering, when $hL \gg 1$ (where L is the distance between the points where the radiation enters and leaves the medium) one can use the diffusion approximation of radiation-transport theory, and in the first approximation, one gets

$$\Gamma_m = N\overline{\Gamma}, \tag{5.15}$$

where N is the mean scattering multiplicity at the observation point and $\overline{\Gamma}$ is the value of the spectral broadening in a single scattering event averaged over a single scattering indicatrix [386] (see also Eq. (5.44)). Of course, the last expression represents only the main part of the asymptotic expansion for high N, but even in this simplified form the diffusion model can answer many questions if N is correctly calculated.

The problem analyzed, within the limits of these assumptions, is identical to the elementary stochastic problem of random walk, the diffusion problem [469]. This identity gives us a serious foundation for applying the method of the random walk of photons in an unlimited scattering medium to describe multiple scattering [386, 390–394]. Keeping in mind that the scattering multiplicity is, in effect, proportional to time, we can use Einstein's famous formula ([290], pp. 229–230)

$$L^2 = 6D_p N. \tag{5.16}$$

The quantity D_p (the mean square of the length of the random step) represents the product of the diffusion coefficient of the photon on the mean time of a photon's free flight. We should mention that D_p and D have a different physical meaning: D_p characterizes the diffusion of photons and D the diffusion of suspended particles.

To calculate D_p, we make use of the fact that the trajectory of a random walk has a profound similarity to a freely coupled polymer chain [470–472]. To calculate the mean scattering multiplicity, we can use the results already utilized to solve problems of configurational statistics of polymer chains [470, 473]. Employing such analogy, it can be suggested that the length of a photon free path corresponds to the length of a link in the chain, while the scattering angle θ_i corresponds to the coupling angle between the links of the chain. However, not only the mentioned analogy should be kept in mind but also the difference between these models. Firstly, the coupling angle between the links of the polymer chain is either completely random (free coupling chain) or strongly fixed, while the scattering angle is a random variable determined according to the scattering indicatrix. Secondly, the length of the free path of the photon is not fixed but distributed according to the exponential law. It can be shown that due to the independence of the scattering angles it is sufficient to replace the term $\mu \equiv \cos\theta$ with its average value $\overline{\mu}$ in the solution to the polymer chain problem (the averaging is performed over a scattering indicatrix in analogy to Eq. (5.44)). In order to account for fluctuations of the length of the free path, according to [473], we can calculate the quantity

$$L^2 = \vec{L}\vec{L} = \left\langle \sum_{i=1}^{N} \sum_{j=1}^{N} \vec{l_i}\vec{l_j} \right\rangle, \tag{5.17}$$

where $\vec{l_i}$ is the vector of translation of the photon between two successive scattering acts. Then in Eq. (5.17), we can perform the following substitution [390]:

$$\vec{l_i} = l\frac{\vec{l_i}}{|\vec{l_i}|}(1 + \Delta_i), \tag{5.18}$$

where l is the mean length of the free path of the photon

$$l = h^{-1}, \tag{5.19}$$

and Δ_i the fluctuating quantity ($\langle \Delta_i \rangle = 0$). The exponential distribution law for $|\vec{l_i}|$ leads to

$$\langle \Delta_i^2 \rangle = 1. \tag{5.20}$$

The products $\Delta_i \Delta_j$ when averaged give a nonzero contribution only when the summation indices coincide ($i = j$), and also double the "diagonal" terms of the sum. Consequently, a term of the type Nl^2 is added to the result obtained for the polymer chain [470, 473], after which we finally obtain [390]

$$L^2 = Nl^2\frac{1+\bar{\mu}}{1-\bar{\mu}} - 2\bar{\mu}l^2\frac{1-\bar{\mu}^N}{1-\bar{\mu}^2} + Nl^2 = Nl^2\frac{2}{1-\bar{\mu}} + o(N). \tag{5.21}$$

Comparing Eqs. (5.21) and (5.16) leads us to the expression [390]

$$D_p = \frac{l^2}{6}\frac{2}{1-\bar{\mu}}, \tag{5.22}$$

which is also well known from the theory of neutron transport [474]. This is not surprising as both neutron [474–477] and radiation [347–349, 354–356, 478] transport theories are based, naturally, on the same principles.

Finally, from Eq. (5.21) we obtain a simple formula to calculate the mean scattering multiplicity

$$N = \frac{1-\bar{\mu}}{2}\left(\frac{L}{l}\right)^2. \tag{5.23}$$

From Eqs. (5.15) and (5.23) we can now get [390]

$$\Gamma_m = \frac{1-\bar{\mu}}{2}\left(\frac{L}{l}\right)^2 \bar{\Gamma}. \tag{5.24}$$

Equations (5.23) and (5.24) are the principle result of the simplified photon diffusion model in an infinite scattering medium. Taking into account Eqs. (5.10) and (5.24) yields

$$\Gamma_m = \frac{1 - \overline{\mu}}{2} \left(\frac{L}{l}\right)^2 \Gamma(90°)(1 - \overline{\mu}) = \frac{1}{2} \left(\frac{L}{l}\right)^2 \Gamma(90°)(1 - \overline{\mu})^2. \qquad (5.25)$$

Let us now analyze Eq. (5.25).

- In conditions of cylindrical geometry, where $L = d\cos(\phi/2)$, we immediately get $\Gamma_m \sim \cos^2(\phi/2)$. This was the result obtained for scattering on model systems (see Figs. 5.7–5.11).

- Furthermore, we can note the proportionality of Γ_m to the square optical thickness, i.e., $\Gamma_m \sim (L/l)^2$ (see Section 5.3).

  In addition, in accordance with Eqs. (5.23) and (5.19), the mean scattering multiplicity in turbid media is proportional to the square of the extinction coefficient. In media with totally chaotically distributed suspended particles the scattering cross sections are additive and, in accordance with Eq. (4.10), we can use the correlation

$$h = \rho\sigma = \frac{3Q}{4r_0}c, \qquad (5.26)$$

where ρ is the number of suspended particles in a unit of volume and σ the complete scattering cross section. Moreover

$$\sigma = Q\pi r_0^2 \qquad (5.27)$$

holds, where Q is the factor of the scattering efficiency [429, 430], r_0 the radius of the suspended particle where scattering takes place, and

$$c = \frac{4}{3}\pi r_0^3 \rho \qquad (5.28)$$

is the volume part of the particles suspended in a dispersed system. Note that in dispersed systems with a developed structure, for example, in gels, Eq. (5.26) can be violated.

- Equation (5.26), taking into account Eq. (5.19), allows us to reformulate the just-arrived conclusion in the following way: in turbid dispersed systems the mean scattering multiplicity is proportional to the square of the concentration of suspended particles. This result confirms, theoretically, the discovered experimental fact $\Gamma_m \sim c^2$ (Figs. 5.7–5.10).

- Finally, looking at Eq. (5.25) we can confirm that the developed diffusion model predicts the same dependence on temperature and viscosity for Γ_m as for Γ, i.e., $\Gamma_m \sim (T/\overline{\eta})$ (Section 5.3, Fig. 5.4).

In conclusion, we should note that the numerous attempts to interpret the multiple scattering spectra using the "successive" approximation method, single, double, triple, . . . , and finally, multiple scattering, due to insurmountable mathematical difficulties turned out to hold little promise (see, e.g., the reviews [350, 352]). The mathematical difficulties can be significantly reduced, as was first shown in [386], if we start straight from high multiplicity scattering. This approach allows one to use the diffusion approximation of the radiation-transport theory which can be reduced to the model of random walk of photons in medium [141, 386, 390–394, 436–438].

Naturally, this approach must be applied consistently and without contradictions. For comparison, we can look at [383, 384] where the authors were apparently the first to reject the "successive" approximation method and suggested to calculate Γ_m using Eq. (5.15). However, according to the statistical model, chosen in one of these works [384], and employed for calculating N, the mean scattering multiplicity is proportional not to the square but to the first order of the extinction coefficient and therefore, has the same value for all observation angles. The analysis carried out in [386, 390–394, 436–438] showed that this was due to inconsistent calculation of the extinction of waves of different multiplicities. The authors did not notice that by applying phenomenological statistics they, in fact, ignored extinction. As for [383], extinction is not considered at all. It is typical that the experiment, performed by them and described in these works, confirmed their theoretical suggestions. In fact, this is not at all surprising. Both the theory and experiment in these studies were concerned, as it turned out, with low multiplicity scattering. Apparently the point here was that the external appearance of highly concentrated water suspensions of latex is always more or less like milk. However, some concentrations are high enough for multiple scattering to develop, while others are not (we have already come across a similar situation [432]: the extinction coefficient of 1% latex is $h \sim 0.1$ cm^{-1}, while for milk it is ~ 10 cm^{-1} ([288], p. 287); if special measures are not undertaken then the difference is indistinguishable to the naked eye). This is evident in both the cited works, as it is well demonstrated in Figs. 5.7 and 5.8, where such curves with concentrations ($c = 0.06$ and $c = 0.113\%$, correspondingly) are shown. This is close to what was used in the discussed studies. It is clear that with such small concentrations the scattering multiplicity is small and, consequently, extinction does not play any noticeable role. In general, it is necessary to account for possible extinction at all stages of the calculations of the multiple scattering spectrum. In transport theory, extinction is automatically included, and therefore as a statistics the random walk statistics should be used.

Thus, we can once more conclude that the simplified diffusion model, worked out in connection with the multiple scattering spectrum [386], turned out to be perfectly capable of giving a qualitatively true principle pattern peculiar to this rather difficult phenomenon. It is also clear that, as the spectrum half-width in this case depends on the optical thickness of the system under study, the approximation of the infinite scattering medium, which is the basis of this model, is not adequate for a real situation and should be replaced. In other words, the model has to somehow take into account the presence of a boundary in the scattering medium. This circumstance is a distinctive feature peculiar to high multiplicity scattering. We should not expect some universal solution from general concepts as the problem is too complex. However, it is worth trying to find, at least, some approaches to achieve this task.

With respect to this problem, we shall now develop the foundations of the radiation-transport theory, its diffusion approximation, and the influence of boundaries, following [141, 386, 390–394, 436–438] by necessity briefly but logically consistent.

5.4.2
Mathematical Model of Multiple Scattering

5.4.2.1 Basic Concepts of Radiation-Transport Theory

For multiple scattering on suspended particles, for every scattering act, the intensity of the scattered light is determined by the differential scattering cross section

$$\sigma_d(\theta) = \frac{1}{4\pi}\sigma p(\theta). \tag{5.29}$$

The phase function $p(\theta)$ is normalized by the condition

$$\frac{1}{4\pi}\int\limits_{4\pi} p(\theta)d\Omega = \frac{1}{2}\int\limits_{0}^{\pi} p(\theta)\sin\theta\,d\theta = 1. \tag{5.30}$$

Radiation propagation in a turbid medium is accompanied by an exponential weakening of its intensity (Bouguer's law). Taking this effect into account, we obtain the following expression for the intensity of single scattering of radiation, I_1

$$I_1 = I_0\frac{\sigma}{4\pi r^2}p(\theta)\exp(-hr), \tag{5.31}$$

where I_0 is the intensity of the external radiation on the boundary of the scattering medium (in the case of single scattering, I_0 corresponds to the radiation intensity incident on a suspended particle), r is the distance from this particle to the observation points.

Single scattered radiation, propagated in a disperse medium, scatters on suspended particles, and high multiplicity scattered radiation arises. At the observation points the intensities of light waves are summed up, running along all possible chains of the scatterers. The consistently applied formula (5.31) leads to the following expression for complete intensity of scattered radiation, I_N:

$$I_N = I_0\sum_j \exp(-hl'_0)\prod_{i=1}^{N_j}\left[\frac{\sigma_i}{4\pi r_i}p(\theta_i)\exp(-hl_i)\right], \tag{5.32}$$

where l'_0 is the distance from the entrance of the radiation into the scattering medium to the first scattering act, the index j enumerates the trajectory of the radiation's movement (the scatterer's chain), N_j the number of acts for the jth trajectory, l_i the distance between ith and $(i+1)$th scatterers. At the same time, the following vector equality is obviously fulfilled

$$\vec{L} = \vec{l}_0 + \sum_{i=1}^{N_j} \vec{l}_i, \tag{5.33}$$

where the vector $\vec{L}$ joins the entry point of the external radiation in the scattering medium with the exit point of the scattered radiation from this medium.

Mathematically, Eq. (5.32) is equivalent to the radiation-transport equation. The radiation-transport theory is known to be an approximate theory, obtained when certain simplified suggestions are made (see [347–349, 354–356] for more). However, this theory's application is so wide that it encompasses not only all dispersed systems, which occur in technical applications, but also in cases which are infinitely far away from them. We can see this by looking at the list of books which use this equality, in one form or another, [469–477].

The conditions for applicability of the radiation-transport equation can be formulated in terms of the following inequalities [354]:

$$l \gg \lambda, \tag{5.34}$$

and

$$l \gg \frac{r_0^2}{\lambda}, \tag{5.35}$$

where l is the length of the radiation free path in a medium. For the length of the light wave, we have $\lambda < 0.7\,\mu m$, and therefore above-mentioned conditions will be fulfilled even in such a turbid media, where $l \sim 0.1$ mm.

5.4.2.2 Multiple Scattering Spectra Determined via the Radiation-Transport Theory

The spectral properties of radiation, leaving the scattering medium, can be described by two equivalent methods: either by the spectral intensity $I(\omega)$, or by the time correlation function of the electrical field, $G(\bar{\tau})$

$$\left| G(\bar{\tau}) \right|^2 = \left\langle I(t+\bar{\tau})\,I(t) \right\rangle - \left\langle I(t+\bar{\tau}) \right\rangle \left\langle I(t) \right\rangle. \tag{5.36}$$

In our opinion, to describe the spectral properties in terms of correlation functions is a more convenient approach due to the following circumstances. Firstly, from the point of view of traditional spectroscopy, the spectrum broadening, caused by scattering on suspended particles, is exceptionally small and to use traditional equipment (spectrometers) here to measure the spectral intensity, $I(\omega)$, is impossible. Therefore, when studying dispersed systems, such technical equipment (correlators) is used which measures directly the function $G(\bar{\tau})$. Secondly, as we have already seen, it is easier to use the language of correlation functions to describe multiple scattering, as in this case the outcome is determined by multiplying correlation functions which characterize separate scattering acts (see, e.g., [411]).

Earlier, when considering radiation-transport theory, we obtained Eq. (5.32) for the total intensity of scattered light, I. In terms of correlation functions, this theory is easily generalized to spectral problems [386]. As we have already seen (see

Eq. (5.10)), light scattering on separate particles undergoing a Brownian motion in the case of free unconstrained diffusion in a dispersed medium leads to broadening of the radiation spectrum by the value Γ. For single scattered radiation, the generalization of Eq. (5.31) is

$$G_1(\tau) = G_0(\tau) \frac{\sigma}{4\pi r^2} p(\theta) \exp\left[-\left(hr + \Gamma\bar{\tau}\right)\right].$$ (5.37)

Finally, we get the following generalization of Eq. (5.32)

$$G(\tau) = G_0(\tau) \sum_j \exp(-hl'_0) \prod_{i=1}^{N_j} \left\{ \frac{\sigma_i}{4\pi l_i^2} p(\theta_i) \exp\left[-\left(hl_i + \Gamma_i\bar{\tau}\right)\right] \right\}.$$ (5.38)

Here the conditions (5.33) should be fulfilled. Further, the correlation function $G(\bar{\tau})$ can be presented as a product of the following multipliers

$$G(\bar{\tau}) = G_0(\bar{\tau}) \frac{I}{I_0} g(\bar{\tau}),$$ (5.39)

where (I/I_0) is given by Eq. (5.32) and $g(\bar{\tau})$ describes the change in the spectrum during multiple scattering. It follows from Eqs. (5.32), (5.38), and (5.39) that the normalized correlation function $g(\bar{\tau})$ is connected to the spectrum broadening in one scattering act, Γ_i, in the following way

$$g(\bar{\tau}) = \left\langle \exp\left(-\sum_{i=1}^{N_j} \Gamma_i\bar{\tau}\right) \right\rangle,$$ (5.40)

where the brackets $\langle \ldots \rangle$ imply averaging along the trajectory. Now we can write $g(\bar{\tau})$ in the following form [390, 422]:

$$g(\bar{\tau}) = \exp\left(-\Gamma_m\bar{\tau} + \sum_{n=2}^{\infty} \frac{K_n(-\bar{\tau})^n}{n!}\right).$$ (5.41)

From a practical point of view, the value of the first cumulant in this expression, which characterizes the half-width spectrum of scattered radiation (see also Eq. (5.9)), is of main interest. From Eqs. (5.40) and (5.41), it follows that

$$\Gamma_m = \left\langle \sum_{i=1}^{N_j} \Gamma_i \right\rangle.$$ (5.42)

Equation (5.42) connects the measured value of the half-width spectrum of multiple scattering of Γ_m with the fundamental characteristics of the dispersed system (see also Eq. (5.14)). The averaging along different trajectories, prescribed by the right part of Eq. (5.42), is fulfilled by the weight coefficient represented by the summation on the right side of Eq. (5.32). By using Eq. (5.26), we can again confirm that Γ_m depends not only on the mean size of the suspended particles, but also on their concentration.

We should reemphasize that this result has a principle meaning: so far correlation spectroscopy methods have been used when single scattering is predominant, i.e., when the measured quantity coincides with Eq. (5.10). From this it follows directly that a dependence of the measured quantity on the concentration of the suspended particles is absent. Consequently, in conditions of multiple scattering there appears a principle possibility to extract, from the obtained value of Γ_m, information not only concerning particle size in dispersed systems, which a single scattering regime also permits, but also concerning its concentration. The practical implementation of this principle possibility is set out in [479,480] (for details, see also Appendix A.2).

5.4.2.3 Transition to High Multiplicity Scattering

From a formal point of view, the mathematical model of multiple scattering can be considered to be completely developed. The obtained expression (5.42), taking into account Eq. (5.32), unambiguously determines the measured value of Γ_m. However, such a most general expression for Γ_m is practically useless as there does not exist an analytical method which allows us to calculate the average on the right side of Eq. (5.42) for arbitrary scattering multiplicities. If such a method existed, it would mean that one could build a universal analytical solution to the radiation-transport equation. At this time, such an universal solution does not exist and, with high probability, will not appear in the foreseeable future.

There are lots of reasons for such a pessimistic outlook. In recent decades the transport equation has been subjected to intensive study due to the fact that the calculations for nuclear reactors are based on the transport theory of neutrons. However, there still has not been even the slightest hint for a possibility of building a universal analytical solution to this equation. Therefore, for such a physical phenomenon as multiple scattering to be used technically, simplified mathematical models must be worked out. Such models should satisfy two basic demands. Firstly, they should allow one analytical calculations to be made, for at least those situations which are of practical interest; secondly, the precision of the solution should satisfy practical requirements. Naturally, such a simplification will be paid for by narrowing of the areas where these models can be applied.

As we have seen above, significant mathematical simplification can be achieved if one immediately goes over to high multiplicity scattering. In particular, in [386] it was suggested to replace Eq. (5.42) by a simpler expression, Eq. (5.15). Here, we can rewrite Eq. (5.15) in a more precise form as

$$\Gamma_m = \langle N \rangle \overline{\Gamma}. \tag{5.43}$$

Here $\langle N \rangle$ is the result of averaging the scattering multiplicity N_j, while averaging along the trajectory leads to weight coefficients which are represented by the summation symbol on the right-hand side of Eq. (5.32). $\overline{\Gamma}$, as usual, is the value of spectral broadening in a single scattering event averaged over a single scattering indicatrix. Taking into account Eqs. (5.10) and (5.30), we obtain

$$\overline{\Gamma} = \frac{1}{4\pi} \int_{4\pi} \Gamma p(\theta) d\Omega = 2D \left(\frac{2\pi n}{\lambda} \right)^2 (1 - \overline{\mu}) = \Gamma(90°)(1 - \overline{\mu}), \tag{5.44}$$

where $\overline{\mu}$ is the result of a similar averaging of the value $\mu = \cos\theta$.

In general, it is clear that the right side of Eq. (5.43) is the main term of the asymptotic expansion of the right part of Eq. (5.42) for $\langle N \rangle \to \infty$. However, for practical applications of correlation spectroscopy of multiple scattering to dispersed systems we need to understand the possible corrections, the next terms of this asymptotic expansion. Let us discuss the procedure for simplifying Eq. (5.42) in more detail taking into account that what follows does not have any relation to critical opalescence itself.

First of all, by using Eqs. (5.10) and (5.44), we can write the identity

$$\sum_{i=1}^{N_j} \Gamma_i \equiv N_j \overline{\Gamma} + \overline{\Gamma} \sum_{i=1}^{N_j} \frac{\overline{\mu} - \cos \theta_i}{1 - \overline{\mu}}. \tag{5.45}$$

With the help of Eq. (5.45), we can rewrite Eq. (5.42) in the following way:

$$\Gamma_m = \langle N \rangle \overline{\Gamma} + C \overline{\Gamma}, \tag{5.46}$$

where

$$C = \frac{1}{1 - \overline{\mu}} \left\langle \sum_{i=1}^{N_j} (\overline{\mu} - \cos \theta_i) \right\rangle. \tag{5.47}$$

In contrast to the approximate expression (5.43), Eq. (5.46) is an exact consequence of Eq. (5.42). The basic problem that arises here is the estimation of the value of C.

Note that C depends on many factors, including, for example, the mean scattering multiplicity of $\langle N \rangle$. Let us introduce the notation

$$\Delta \mu_{iN} \equiv \overline{\mu} - \langle \cos \theta_i \rangle_N, \tag{5.48}$$

where the index N signifies averaging only along those trajectories which are made up of the given number of links of N. If, as done in [386], we look at the diffusion of photons in an infinite scattering medium and one end of the trajectory is fixed, while the second one is free, then in this model all angles θ_i will be independent random quantities with the same distribution function, $p(\theta)$. It could then be assumed that $\Delta \mu_{iN} = 0$, and therefore $C = 0$. In fact, the real situation is significantly more complex. The second end of the trajectory is also fixed, which is expressed by the condition (5.33). This, in turn, means that the distribution of random quantities can differ from the phase function $p(\theta)$. When N has a large value one would expect that for links in the middle of the trajectory the shown difference will be insignificant, i.e., the corresponding $\Delta \mu_{iN}$ will be almost equal to zero. Nonzero values can be expected only at the ends of the trajectories. Increasing the distance from the ends of the trajectory, the values $\Delta \mu_{iN}$ will apparently constantly tend to zero. Therefore, it can be assumed that only a limited number of "end" terms will play a substantial role in the sum on the right side of Eq. (5.47). As a result, we can state that when $\langle N \rangle \to \infty$

$$C = O(1). \tag{5.49}$$

This result, in particular, has the consequence that Eq. (5.43) really describes the main term of the asymptotic expansion of Γ_m for high multiplicity scattering. In accordance with Eq. (5.49) it can be suggested that in conditions of geometric similarity when $\langle N \rangle \to \infty$ the value of C tends to a finite threshold value, Γ_0. Thus for high mean scattering multiplicities, the exact formula (5.46) can be replaced by a more convenient approximated expression

$$\Gamma_m \approx \langle N \rangle \, \overline{\Gamma} + C_0 \overline{\Gamma}, \tag{5.50}$$

where the constant C_0, seemingly, depends on the form of the phase function, $p(\theta)$. Equation (5.50) is an improvement as compared to the original formula (5.43). It leads to the following principle conclusion: in turbid media, Γ_m is a linear function of the mean scattering multiplicity. This was assumed from the very beginning, but now this result has been obtained as a consequence of the diffusion approximation in the theory of radiation transport. This result also shows that the nondiffusive contribution, represented in the tern Γ_0, really plays a role.

Only empirical data exist regarding the possible values of the coefficient C_0 (see Section 5.3 and Fig. 5.14). They support the fact that in technical applications the second term in the right side of Eq. (5.50) cannot be neglected. However, the simple dependence between Γ_0 and the size of the scatterers (see Fig. 5.14) [393, 394] gives some hope that before a final theoretical solution to the problem is discovered, it might be solved for practical applications, at least, empirically.

Let us now look at the exact expression (5.46). It is clear that trajectory averaging depends on the experiment's geometry and on the mean length of the free path l, defined using Eq. (5.19) (see also Eq. (5.26)). The value of the diffusion coefficient of suspended particles D in no way affects this averaging. Therefore $\langle N \rangle$ and C do not depend on such typical characteristics of the medium as temperature, T, and viscosity, η. Among the quantities on the right-hand side of Eq. (5.46) only $\overline{\Gamma}$ depends on T and $\overline{\eta}$. By looking at Eqs. (5.44) and (5.10) together, we can be sure that, in full accordance with experimental results (see Section 5.3), Γ_m is proportional to absolute temperature and inversely proportional

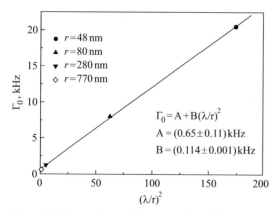

Figure 5.14 Dependence of Γ_0 on the size of the scatterers [393, 394].

to the viscosity of the dispersed medium, like the spectrum half-width of single scattering.

5.4.2.4 Effect of the Shape of the Sample on the Mean Scattering Multiplicity

The diffusion model, described in Section 5.4, is attractive because occasionally it leads to a simple expression for mean multiplicity of scattering. The main shortcoming of this simplified model is that, as we have already shown, the influence of the boundary of the scattering medium on the statistics of the random walk of scattered radiation is neglected. In order to determine the medium scattering multiplicity accurately, with respect to technological applications, one needs to go over sufficiently to a more complex multiple scattering model which is based on solving the boundary value problems for the differential equation describing the diffusion of photons in a scattering medium [390–394, 437, 481].

In the proposed mathematical model here, the well-known equation

$$D\nabla^2 \rho\left(\vec{r}, t\right) = \frac{\partial \rho\left(\vec{r}, t\right)}{\partial t} \tag{5.51}$$

is used [482], where D is the diffusion coefficient, $\rho(\vec{r}, t)$ the density of the diffusing particles at the point, $\vec{r}$, at the moment of time, t. It should be mentioned that such kind of equations is usually denoted as thermal conductivity equations and their properties are studied in detail in the corresponding literature (see, e.g., [483]). In particular, the well-known solution to this equation for an impulse source in an infinite medium leads to Eq. (5.52)

$$\langle |\Delta\vec{r}|^2 \rangle = 6Dt, \tag{5.52}$$

which is analogous to Eq. (5.16). In Eq. (5.51), $\Delta\vec{r}$ is the displacement of the diffusing particles.

The diffusion equation can also be applied to describe the intensity distribution of scattered radiation with respect to the scattering multiplicity. Let us suppose that at the initial moment of time, the pulse input of the external radiation occurs in the scattering medium. Such nonstationary problems are described by Eq. (5.51). However, it is clear that in this case, as was mentioned above, the scattering multiplicity N is directly proportional to time, t. This means that by replacing the variable t by N, we keep the form of Eq. (5.51) and this only influences the value of the constant coefficient on the left-hand side. If we compare Eqs. (5.52) and (5.16) we come to the conclusion that the new value for the shown coefficient corresponds to the value of D_p. As a result, by using Eq. (5.22), which was obtained earlier, we get the differential equation [437]

$$\frac{l^2}{3(1 - \bar{\mu})}\nabla^2 I(\vec{r}, N) = \frac{\partial I(\vec{r}, N)}{\partial N}, \tag{5.53}$$

where $I(\vec{r}, N)$ is the intensity of N-multiple scattering. It should be noted that $I(\vec{r}, N)$ is continuous in N since the function $I(\vec{r}, t)$ is continuous.

The mathematical model of multiple scattering under consideration is based on Eq. (5.53). However, for a complete formulation of the mathematical problem it is necessary to formulate the initial and boundary conditions. By integrating both parts of Eq. (5.53) with respect to N, we obtain an equation which should clearly coincide with the well-known [354, 355] diffusion equation for the complete intensity of scattered radiation. These requirements are satisfied by the initial condition

$$I(\vec{r}, 0) = \frac{1}{(1 - \bar{\mu})} I_0(\vec{r}), \tag{5.54}$$

where $I_0(\vec{r})$ has the previous meaning of the intensity of coherent external radiation, entering the scattering medium through its boundary without experiencing so far a single scattering act. The boundary conditions for Eq. (5.53) are formulated similarly to the boundary conditions for the well-known radiation diffusion equation [355]. It is convenient to formulate them as first-order conditions on the "displaced" boundary

$$I(\vec{r} + z_0 \vec{e}, N) = 0, \tag{5.55}$$

where $\vec{r}$ is any of the points on the medium's boundary where is no external radiation, $\vec{e}$ is the external boundary normal at this point, and z_0, the so-called extrapolated length, is determined in the following way (see [467]):

$$z_0 = \frac{2l}{3(1 - \bar{\mu})} \approx \frac{0.7l}{1 - \bar{\mu}}. \tag{5.56}$$

The mean scattering multiplicity is as usual determined by

$$\langle N \rangle = \frac{\displaystyle\int_0^\infty N I(\vec{r}, N) dN}{\displaystyle\int_0^\infty I(\vec{r}, N) dN}, \tag{5.57}$$

where the solution of the boundary value problem obtained from Eqs. (5.53) to (5.55) should be substituted. Naturally, the use of Eqs. (5.53)–(5.55) is limited by the applicability of the conditions of the diffusion approximation [354]. Radiation absorption in turbid media should be negligible and the typical size of the scattering sample L should satisfy the inequality

$$L \gg z_0. \tag{5.58}$$

Now we can have a detailed look at the general properties of the solution of the boundary value problem given by Eqs. (5.53)–(5.55), which arise from the symmetry of Eq. (5.53) with respect to the similarity transformation. We shall, as previously, consider L to be the distance between the entry point of the coherent external radiation in the scattering medium and the exit point of scattered radiation from this medium. The symmetry of Eq. (5.53) allows us to express the solution of the original problem using the solution of geometrically similar problems [440]

$$I(\vec{r}, N) = \frac{1}{L^2} I_1\left(\frac{\vec{r}}{L}, \frac{N}{L^2}\right), \tag{5.59}$$

where $I_1(\vec{r}, N)$ is the solution of the geometrical similar problem for the unit value of L (to be more precise, as a similarity coefficient we should use the value $L + 2z_0$, however, due to inequality (5.58) in a first approximation this can be ignored). As a result, by inserting Eq. (5.59) into Eq. (5.57), we can conclude that the quantity $\langle N \rangle$ is proportional to the square of (L/l). This result can be expressed via

$$\langle N \rangle_d = \frac{1 - \overline{\mu}}{2} \left(\frac{L}{l}\right)^2 F, \tag{5.60}$$

where the subscript d signifies the "diffusion model," Eqs. (5.53)–(5.55).

In comparison to Eq. (5.53), the result obtained for the previously discussed simplified diffusion model, this formula contains the extra multiplier, F, depending on the shape of the scattering sample [437]. In fact, we managed to get rid of the influence of this multiplier by going over from the representation of the obtained experimental dependences in form of the graphs shown in Figs. 5.8 and 5.11 to those shown in Figs. 5.9 and 5.12.

The analytical determination of the mean scattering multiplicity $\langle N \rangle_d$, which, as expected, was only possible for certain problems with a high degree of symmetry of the scattering sample, which unfortunately does not include cylindrical geometry [440, 481]. Solutions are obtained for the following cases:

- The scattering medium fills a sphere, in the center of which is the coherent radiation source. In this case, for radiation exiting through the spherical boundary, $F = 1$ holds.

- The scattering medium fills a half sphere, while the narrow beam of coherent radiation is directed at the center of its planar boundary. In this case, $F = 0.6$ holds on the spherical boundary.

- On the planar boundary, the value of F changes from 0.6 close to the external edge to 3 in its central part.

All these examples examined by Pavlov [440] show that the form of the scattering sample substantially influences the value of the medium scattering multiplicity. Consequently, any model which does not take into account the form of the scattering sample will be unsatisfactory from a practical point of view.

Thus, although for particular examples of geometry with a high degree of symmetry it is possible to calculate the factor F analytically, it is impossible to find a solution to this problem, in general, i.e., with an arbitrary shape of the scattering sample. Therefore, in any concrete case the coefficient F should be determined by the calibration measurement on the dispersed systems with known values of $\overline{\Gamma}, \overline{\mu}$, and l. In our opinion, an alternative method of determination of F could consist in the application of the Monte Carlo method (see, e.g., [484]).

In order to complete the construction of the mathematical model, we have to substitute Eq. (5.60) into Eq. (5.50) which describes the spectrum half-width of multiple scattering, Γ_m. Note that the value of $\langle N \rangle_d$ is an approximation of the mean scattering multiplicity of $\langle N \rangle$. It can be assumed that with high scattering multiplicity the difference in these values approaches the constant value N_0. As a result, Eq. (5.50) leads to the expression

$$\Gamma_m = \frac{1}{2}(1 - \overline{\mu}) F \left(\frac{L}{l}\right)^2 \overline{\Gamma} + (C_0 + N_0)\overline{\Gamma}. \qquad (5.61)$$

This formula allows us to verify again the basic conclusion derived in [141, 386, 390–394, 436–438, 440, 481] by theoretical and experimental research: in turbid media with geometrically similar conditions the measured spectrum broadening of Γ_m is a linear function of the square of the optical thickness of the scattering sample.

5.5
On the Nature of the Constant Γ_0

5.5.1
The Relation of Γ_0 to the Size of the Scatterers

The question concerning the relation of Γ_0 with the scatterer's size is extremely interesting, not only for the theory and practice of spectral research in optically dense dispersed systems but it becomes especially important when studying critical phenomena. As shown above, all the different mathematical models, which were based on the diffusion approximation theory of radiation transport (see Eqs. (5.46), (5.50), and (5.61)), predicted its existence. Although the physical nature of Γ_0 is unclear, experimental data concerning multiple scattering in latexes of different sizes, which have already been presented (see Figs. 5.7–5.9, 5.11–5.13), allow us to suggest an empirical dependence of Γ_0 on the size of the scatterer. The results for four latex particle sizes are shown in Fig. 5.14 [393, 394]. The data presented make it clear that the correction Γ_0 depends neither on the size of the cuvette nor on the concentration of the dispersed system. Further, it turned out [392] that we can observe a dependence of Γ_0 on the polarization of the registered radiation. All these facts allowed us to make the suggestion about the nondiffusion character of this contribution [392].

While this problem is extremely important and interesting for multiple light scattering on Brownian particles, as can be seen in Fig. 5.14, it does not at all touch on critical opalescence. In fact critical opalescence is a sharp strengthening of light scattering, mostly forward, and arises close to the critical point. It begins when the correlation radius, ξ, and the wave length of probe radiation, λ, become comparable, i.e. $(\lambda/\xi)^2 = (\lambda/r)^2 \leq 1$. As the increase of ξ does not stop (see Eq. (4.24)) when approaching the critical point, it is clear that under conditions of developed critical opalescence the value of Γ_0 is close to zero (see Fig. 5.14) and consequently, the role of this contribution in the spectrum half-width of multiple scattering on critical fluctuations is insignificant in comparison to scattering on

Brownian particles. This result is a principle conclusion for critical opalescence (see Chapter 6). It seems that further theoretical and experimental efforts are necessary to explain the nature of this term.

5.5.2
The Relation of Γ_0 to the Depth of the Diffusion Source

Recently efforts have been undertaken to establish the connection between Γ_0 and the depth of the diffusion source. In [485, 486] the distribution of the intensity of diffused radiation on the surface of the spherical cuvette, filled with a multiple scattering medium, was investigated. The result was quite unexpected: in the case of developed multiple scattering the equivalent source is located considerably deeper than that as suggested by Eq. (5.56).

Thus, one cannot exclude the possibility that the correction to Γ_0 could appear even within the framework of the diffusion approximation. This is because the solution of the diffusion equation with variable distance between the effective source and the medium's boundary leads to significant changes in the values of the registered spectrum half-width [485, 486]. For deeper understanding of the nature of Γ_0 and also of other peculiarities of multiple scattering, investigations into this field have to be continued.

6

Critical Opalescence: Theory and Experiment

6.1
Introduction

As was shown in the previous chapter, the results of investigation of the key features of the multiple scattering spectra in model systems (concentrated latexes) were successfully explained on the basis of the diffusion approximation of the radiation-transport theory [141, 386, 390–394, 436–438, 440, 481].

In [387–389], the dependences obtained in model systems, were extended to critical fluctuations. In model systems the extinction coefficient is also temperature independent due to the stability of the dispersed phase particles with respect to temperature changes. Therefore, the temperature dependence of the multiple scattering spectrum half-width, Γ_m, as was shown in Section 5.3, will be the same as for single scattering.

A qualitatively different situation arises for light scattering on density (concentration) fluctuations close to the critical point. Here, on the contrary, when $T \rightarrow T_c$ along the critical isochore the rapid growth in the fluctuation intensity is accompanied by an increase in extinction (decrease in the photon mean free path) and a completely different temperature dependence of Γ_m should be expected.

In order to determine the kind of dependence it is, the results obtained for dispersed systems with the help of the simplified diffusion model (Eq. (5.24)) can be used as a basis. It should be mentioned that the more general formula, which refines this model, includes the factor F, taking into account the experiment's geometry, and also an additional term (Γ_0), which depends neither on the scattering medium's geometry nor on the scatterer's concentration (see Eq. (5.61)). As the geometric factor, F, does not depend on temperature, in the case of critical opalescence it can be simply omitted.

As for the extra contribution (Γ_0) in the half-width spectrum its nature, as shown in Section 5.5, is such that when the size of the scatterers grows its value decreases but when the scatterers reach sizes equal to λ, Γ_0 almost completely disappears (see Fig. 5.14) [393, 394]. Therefore, as the critical point is approached the contribution of Γ_0 in the half-width spectrum should decrease, as discussed Chapter 5, and the half-width spectrum temperature dependence will be mainly determined by the diffusion component (5.24).

Critical Behavior of Nonideal Systems. Dmitry Yu. Ivanov
Copyright © 2008 WILEY-VCH Verlag GmbH & Co. KGaA, Weinheim
ISBN: 978-3-527-40658-6

6.2
Theory of Critical Opalescence Spectra

6.2.1
Analysis of the Behavior of Γ_m Close to the Critical Point

According to Eq. (5.24), for high scattering multiplicity the temperature dependence of Γ_m is determined by the behavior of three quantities: l, $\overline{\Gamma}$, and $1 - \overline{\mu}$. When $T \to T_c$ all these quantities monotonously decrease. In fact, fluctuation growth leads to an increase in extinction and a decrease in l ($hl = 1$). At the same time, the correlation radius ξ grows and consequently the scattering indicatrix extends in the forward direction and $\overline{\Gamma}$ and $1 - \overline{\mu}$ decrease. According to the same formula (5.24) a decrease in l creates a precondition for the growth of Γ_m. However, a simultaneous decrease in $\overline{\Gamma}$ and $1 - \overline{\mu}$ can lead to a directly contradictory result. Therefore, for further analysis, it is necessary to clarify which of these tendencies will be predominantly close to the critical point.

In order to discuss critical opalescence it is more convenient to rewrite the basic formula, Eq. (5.24), as follows:

$$\Gamma_m = \frac{1}{2} L^2 \frac{1}{l_{tr}} \frac{\overline{\Gamma}}{l},$$ (6.1)

where the temperature dependence is included into two factors, $(1/l_{tr})$ and $(\overline{\Gamma}/l)$, and l_{tr} is given by

$$l_{tr} = \frac{l}{1 - \overline{\mu}}.$$ (6.2)

The transport mean free path, l_{tr}, is the distance the particle covers in the direction of the initial movement in the scattering medium. In transport theory, l_{tr} generalizes the notion of the mean free path in media with noticeable scattering anisotropy ([291], p. 681).

Averaging the quantities in Eq. (6.1) over scattering angles has to be performed with a weighting function proportional to dh, which is the differential extinction coefficient (see Eqs. (4.10), (4.11), (4.15), and (5.19)) Since according to the definition

$$\frac{1}{l} = \int dh$$ (6.3)

holds, then $1 - \overline{\mu}$ and $\overline{\Gamma}$ can be obtained from Eqs. (6.4) and (6.5), respectively,

$$\frac{1}{l_{tr}} = \int (1 - \mu)\, dh,$$ (6.4)

$$\frac{\overline{\Gamma}}{l} = \int \Gamma\, dh.$$ (6.5)

Thus, to find the temperature dependence of Γ_m it is sufficient to study the behavior of the latter two integrals [387, 389].

Let us start with the qualitative analysis of the temperature dependence of the integral given by Eq. (6.4). The differential extinction coefficient, dh, characterizes scattering in the solid angle $d\Omega$. For $T \to T_c$, the quantity $(dh/d\Omega)$ should grow continuously for any scattering angle. Limiting (maximum) values of $(dh/d\Omega)$ are reached when $T = T_c$. It is known that at the critical point $(dh/d\Omega)$ diverges when scattering is in the forward direction $(\theta = 0)$ and has a finite value for any other scattering angle (see, e.g. [140]). The limiting values of the integrand in Eq. (6.4) are finite for all scattering angles, as the singularity at $\theta = 0$ disappears when $(dh/d\Omega)$ is multiplied by $(1 - \cos\theta)$. As a result it becomes clear that when $T \to T_c$ the integral grows continuously approaching a finite limiting value. It should be mentioned that the authors of [345] arrived at qualitatively similar conclusions in their investigation in the extinction coefficient behavior for multiple light scattering close to the critical point on the basis of the Dyson and Bethe–Salpeter exact equations [487].

When $T \to T_c$ the integral equation (6.5) behaves in a similar way. The increase of $(dh/d\Omega)$ outweighs the decrease in Γ. Continuously increasing integrand tends toward a finite limit: the singularity at $\theta = 0$ disappears due to multiplication by

$$\Gamma \propto q^{3+x_{\overline{\eta}}} \propto (1 - \cos\theta)^{(3+x_{\overline{\eta}})/2} \tag{6.6}$$

(see Eqs. (4.3), (4.30), and (4.31)). Thus, we can formulate the main conclusion: as the system approaches the critical point the multiple scattering spectrum half-width grows monotonically, reaching its maximum value, Γ_{max}, directly in the critical point. For comparison, it can be noted that the half-width of the single scattering spectrum Γ when $T \to T_c$, on the contrary, decreases [26, 306–308, 319]. Moreover, $\Gamma_m \sim L^2$ holds [386, 390, 392], while Γ is completely independent of the scattering system dimensions.

In order to clarify how exactly the multiple scattering spectrum half-width grows and what its maximum value is, it is necessary not only to analyze the character of qualitative changes of the Γ_m-temperature dependence, but also to find out what analytical form it takes. Taking into account Eqs. (4.1), (4.9), and (4.10), the differential extinction coefficient close to the critical point can be written as

$$dh = \frac{8\pi^3}{3\lambda^4}\left(m\frac{\partial n^2}{\partial m}\right)^2 k_B T \chi C_1 \left(1 + q^2\xi^2\right)^{-1+(\eta/2)} I(\theta, \psi)\, d\Omega. \tag{6.7}$$

Here, m is a variable that defines the order parameter, i.e., density ρ for pure liquids and concentration c for binary mixtures, χ is the susceptibility, which as always is given by

$$\chi = \chi_0 \tau^{-\gamma}. \tag{6.8}$$

As for the constant C_1, it is known that for all possible versions of scaling representation of the correlation function, connected to the structural factor, its value lies within the limits $C_1 = 0.9 - 1.0$ [57, 334, 371, 488].

The indicatrix of dipole scattering, $I(\theta, \psi)$ in Eq. (6.7), equals

$$I(\theta, \psi) = \frac{3}{8\pi}\sin^2\Phi = \frac{3}{8\pi}\left(1 - \sin^2\theta\cos^2\psi\right), \tag{6.9}$$

where ψ is the angle between the direction of the vector $\vec{E}$ in the incident wave and the projection of the scattering vector on the plane, perpendicular to the propagation direction. In Eq. (6.9) it is taken into account that for the chosen angles $\cos \Phi = \sin \theta \cos \psi$ holds. The indicatrix is normalized in such a way that the integral from Eq. (6.9) over the full solid angle is equal to unity [284, 288].

To calculate the integrals (6.4) and (6.5), the indicatrix $I(\theta, \psi)$ is multiplied by functions which are independent of the azimuthal angle ψ. Therefore, it can be immediately averaged over ψ and Eq. (6.10)

$$I(\theta) = \frac{3}{16\pi} \left(1 + \cos^2 \theta\right) \tag{6.10}$$

henceforward can be used in place of Eq. (6.9) in all formulae. For further analysis, Eq. (6.7) accounting for Eqs. (4.3), (4.24), and (6.8) can be transformed into the more convenient form

$$dh = h_0 \frac{3}{16\pi} \frac{1 + \cos^2 \theta}{\left[\left(\sqrt{2}k\xi\right)^{-2} + 1 - \cos \theta\right]^{1-(\eta/2)}} d\Omega, \tag{6.11}$$

where

$$h_0 \equiv \frac{8\pi^3}{3\lambda^4} \left(m\frac{\partial n^2}{\partial m}\right)^2 C_1 \frac{k_B T \chi_0}{\left[\left(\sqrt{2}k\xi_0\right)^2\right]^{1-(\eta/2)}}. \tag{6.12}$$

Here h_0 is the extinction coefficient in a simple system composed of chaotically positioned dipole scatterers, provided that the single scattering constant at $90°$ is the same as in the critical point.

In addition, the well-known relation (4.35), which exists between the critical indices γ, ν, and η, is accounted for in Eq. (6.12). Thanks to this the explicit dependence on τ in Eq. (6.7) disappears in Eq. (6.12). When Eqs. (6.10) and (6.11) are inserted into Eqs. (6.3)–(6.5), we get integrals of the type

$$J(\zeta, u) = \frac{3}{16\pi} \int (1 + \cos^2 \theta)(2u + 1 - \cos \theta)^{\zeta - 1} d\Omega, \tag{6.13}$$

where $(u)^{-1} = 2\left(\sqrt{2}k\xi\right)^2$.

With the help of the substitution

$$2y = 2u + 1 - \cos \theta, \tag{6.14}$$

the result of integration of Eq. (6.13) for $\zeta > 0$ can be written in a simple, although rather awkward, form as

$$J(\zeta, u) = \frac{3}{2}2^{\zeta}\left[(1 + 2u + 2u^2)\frac{(1+u)^{\zeta} - u^{\zeta}}{2\zeta}\right.$$
$$\left. - (1 + 2u)\frac{(1+u)^{1+\zeta} - u^{1+\zeta}}{1+\zeta} + \frac{(1+u)^{2+\zeta} - u^{2+\zeta}}{2+\zeta}\right]. \tag{6.15}$$

The latter relation makes it possible to analyze the behavior of all the quantities of interest close to the critical point.

6.2.1.1 Calculation of the Limiting Values of Key Quantities [141, 389]

Directly in the critical point, where $q\xi \equiv x \to \infty$, $u = 0$, Eq. (6.15) gets the simple form

$$J(\zeta, u) = J(\zeta) = \frac{3}{2} 2^\zeta \left(\frac{1}{2\zeta} - \frac{1}{1+\zeta} + \frac{1}{2+\zeta} \right). \tag{6.16}$$

It is clear that

$$J(1) = J(2) = 1 \quad \text{and} \quad \forall \zeta \in (1; 2) : J(\zeta) \approx 1. \tag{6.17}$$

It should be noted that $J(1) = 1$ is a direct consequence of normalizing equations (6.9) and (6.10). In the range of small ζ, Eq. (6.16) goes over to

$$J(\zeta) \cong \frac{3}{4\zeta}, \tag{6.18}$$

which satisfies the well-known logarithmic singularity of the extinction coefficient of single scattering in the Ornstein–Zernike approximation ($\eta = \zeta = 0$).

Equations (6.11), (6.13), and (6.16) give the limiting integral values of Eqs. (6.3) and (6.4)

$$\lim_{T \to T_c} \frac{1}{l} = h_0 J \left(\frac{\eta}{2} \right) \equiv \frac{1}{l_0} J \left(\frac{\eta}{2} \right), \tag{6.19}$$

$$\lim_{T \to T_c} \frac{1}{l_{tr}} = h_0 J \left(1 + \frac{\eta}{2} \right) \equiv \frac{1}{l_0} J \left(1 + \frac{\eta}{2} \right), \tag{6.20}$$

where l_0 is understood as a certain characteristic length

$$l_0 \equiv \frac{1}{h_0}. \tag{6.21}$$

We have already extensively discussed (see Section 4.4) the value of the Fisher critical index η (see Eq. (4.8)) and the problems of determining it, while here only the fact that for all calculation methods applied to three-dimensional models it has got values within the range $0.031 < \eta \leq 0.041$ [40] is important. For such values of η the results of calculating the integrals $J(\zeta)$ in Eqs. (6.19) and (6.20), carried out according to Eqs. (6.18) and Eq. (6.17), respectively, differ from the calculation using the exact formula (6.16) by only a few tenths of a percent. We shall, therefore, from now on be guided by these simplified estimates.

To calculate the limiting integral value of Eq. (6.5) we shall use the results obtained in Chapter 4. So the single scattering spectrum half-width is determined by Eq. (4.29) which can be rewritten as follows, taking into account Eqs. (4.31) and (4.32)

$$\Gamma = \Gamma_1 R \frac{(q\xi)^2}{(k\xi)^{3+x_{\overline{\eta}}}} \Omega(q\xi), \tag{6.22}$$

where

$$\Gamma_1 \equiv \frac{k_B T}{6\pi \overline{\eta}_0 \xi_0^3} (k\xi_0)^{3+x_{\overline{\eta}}}. \tag{6.23}$$

When the function suggested by Paladin and Peliti (Eq. (4.33)) is inserted in Eq. (6.22) instead of $\Omega(q\xi)$, and also taking into account Eqs. (4.27) and (4.3), we obtain

$$\lim_{T \to T_c} \Gamma = \Gamma_{\max} = \frac{3\pi\sqrt{2}}{4} 2^{x_{\overline{\eta}}/2} \left(\frac{3\pi}{8}\right)^{-x_{\overline{\eta}}-\eta} R\Gamma_1 (1 - \cos\theta)^{(3+x_{\overline{\eta}})/2} \tag{6.24}$$

$$\cong \frac{3\pi\sqrt{2}}{4} R\Gamma_1 (1 - \cos\theta)^{(3+x_{\overline{\eta}})/2}.$$

Keeping this latter result in mind and taking into account Eqs. (6.11) and (6.21), the limiting integral value of Eq. (6.5) can be represented as

$$\lim_{T \to T_c} \frac{\overline{\Gamma}}{l} \cong \frac{3\pi\sqrt{2}}{4} R \frac{\Gamma_1}{l_0} J\left(\frac{3 + x_{\overline{\eta}} + \eta}{2}\right), \tag{6.25}$$

which, for all possible critical index values in Eq. (6.25), leads to a very simple expression

$$\lim_{T \to T_c} \frac{\overline{\Gamma}}{l} \cong \pi R \frac{\Gamma_1}{l_0}. \tag{6.26}$$

Finally, let us write the formula for the limiting value of Γ_m in the critical point. Considering Eqs. (6.20) and (6.26) and the estimates (6.17) and (6.18), for Eq. (6.1) we get

$$\Gamma_{\max} = \lim_{T \to T_c} \Gamma_m \cong \frac{1}{2} \pi R L^2 \frac{\Gamma_1}{l_0^2}. \tag{6.27}$$

This formula makes it clear that the limiting value of the multiple scattering spectrum half-width at the critical point is completely determined by the combination $\Gamma_1 l_0^{-2}$ or $\Gamma_1 h_0^2$ for the given dimensions of the scattering media and the experiment's geometry (L^2). This makes it possible to analyze the dependence of $\Gamma_{\max}$ from the properties of the studied system, probe radiation and dynamic critical indices. Taking into account Eqs. (6.12) and (6.23) we get [387]

- The dependence of $\Gamma_{\max}$ on ξ_0 in the form

$$\Gamma_{\max} \propto \xi_0^{-4+2\eta+x_{\overline{\eta}}}. \tag{6.28}$$

This result provides a unique possibility to determine a very important characteristic of critical phenomena, the amplitude of the correlation length. We should note that the limiting value of the multiple scattering spectrum half-width is extremely sensitive to the choice of the value ξ_0: a 10% change in the value of ξ_0 leads to a spectacular change of $\Gamma_{\max}$ of nearly one and a half times (!). This result is extremely interesting considering that ξ_0 is one of the toughest values (without counting, of course, the index η) to measure. This dependence's main feature is that it is almost unconnected with the choice of the values of the dynamic critical indices ($2\eta + x_{\overline{\eta}} \ll 1$). This means that it does not lose its strength even if, for one reason or another (see Chapters 1 and 2), the system's critical behavior becomes "classical" in the immediate vicinity of the critical point ($\eta = x_{\overline{\eta}} = 0$).

- *The dependence $\Gamma_{\max}$ on λ in the form*

$$\Gamma_{\max} \propto \lambda^{-7-2\eta-x_{\overline{\eta}}}. \tag{6.29}$$

Here we can observe quite nontrivial dependence of $\Gamma_{\max}$ on the probe radiation wavelength. Unfortunately, it is impossible to obtain the values of the dynamic critical indices from this relation, however this relation is itself still quite remarkable.

- *The dependence of $\Gamma_{\max}$ on $k\xi_0$ in the form*

$$\Gamma_{\max} \propto (k\xi_0)^{-1+2\eta+x_{\overline{\eta}}}. \tag{6.30}$$

This formula shows that even by looking at $k\xi_0$ there is, unfortunately, no chance of determining the dynamic critical indices. However, this dependence gives the unprecedented opportunity to select one or another universal scaling function.

Let us perform this comparison for the Kawasaki, Eq. (4.25), and Paladin–Peliti, Eq. (4.33), relations via Eq. (6.31),

$$\left(\frac{\Gamma_{m,K}}{\Gamma_{m,P}}\right)_{\max} = \lim_{T \to T_c} \frac{\Gamma_{m,K}}{\Gamma_{m,P}} = \frac{\Gamma_{1K} h_{0K}^2}{\Gamma_{1P} h_{0P}^2} \cong (k\xi_0)^{-(2\eta+x_{\overline{\eta}})}. \tag{6.31}$$

It is easy to see that $(\Gamma_{m,K})_{\max}$ exceeds $(\Gamma_{m,P})_{\max}$ by 1.5–2.5 times depending on the various combinations of possible parameter values inserted into Eq. (6.31). This fact shows the principle way of obtaining information on the critical indices η and $x_{\overline{\eta}}$ from experiment in the region of developed multiple scattering.

As Ω_B in Eq. (4.34) and Ω_P in Eq. (4.33), unlike Ω_K, have the same asymptotic form, the result (6.31) remains the same even when Ω_P changes to Ω_B in this formula.

For more convenience we shall close this part by bringing together all the obtained regularities

$$\Gamma_{\max} \propto \xi_0^{-4+2\eta+x_{\overline{\eta}}}, \quad \Gamma_{\max} \propto \lambda^{-7-2\eta-x_{\overline{\eta}}}, \quad \Gamma_{\max} \propto (k\xi_0)^{-1+2\eta+x_{\overline{\eta}}}. \tag{6.32}$$

6.2.1.2 Calculation of the Temperature Dependence [141, 389]

For the sake of simplicity we shall further discuss the temperature behavior of dimensionless quantities which satisfy the relations

$$h^* = \frac{h}{h_0}, \quad l^* = \frac{l}{l_0}, \quad l_{\text{tr}}^* = \frac{l_{\text{tr}}}{l_0}, \quad \Gamma^* = \frac{\Gamma}{\Gamma_1}, \quad \overline{\Gamma}^* = \frac{\overline{\Gamma}}{\Gamma_1}. \tag{6.33}$$

We immediately obtain from Eq. (6.22) the dimensionless single scattering spectrum half-width as

$$\Gamma^* = R \frac{(q\xi)^2}{(k\xi)^{3+x_{\overline{\eta}}}} \Omega (q\xi) .$$ (6.34)

When analyzing the temperature dependence of the multiple scattering spectrum it is convenient to use a dimensionless variable which does not contain the geometric distance L. Let us introduce it as

$$\Gamma_m^* = 2 \frac{\Gamma_m}{\Gamma_1} \frac{l_0^2}{L^2},$$ (6.35)

then the expression (6.1) can be rewritten as

$$\Gamma_m^* = \frac{1}{l_{tr}^*} \frac{\overline{\Gamma^*}}{l^*}.$$ (6.36)

To get the temperature dependence of the dimensionless free path lengths (ordinary and transport) Eq. (6.11) should be substituted in the integrals, Eqs. (6.3) and (6.4). After a simple transformation, and taking into account Eq. (6.15), we get, by analogy with Eqs. (6.19) and (6.20),

$$\frac{1}{l^*} = \int dh^* = J \left(\frac{\eta}{2}, \frac{1}{2 \left(\sqrt{2} k\xi \right)^2} \right),$$ (6.37)

$$\frac{1}{l^*} = \int (1 - \mu) dh^* = J \left(1 + \frac{\eta}{2}, \frac{1}{2 \left(\sqrt{2} k\xi \right)^2} \right) - \frac{1}{\left(\sqrt{2} k\xi \right)^2} J \left(\frac{\eta}{2}, \frac{1}{2 \left(\sqrt{2} k\xi \right)^2} \right).$$ (6.38)

Before analyzing the obtained expressions, let us look at one of the most important special cases, the Ornstein–Zernike approximation ($\eta = 0 \rightarrow \zeta = 0$). As the basic relation (6.15) diverges under these conditions it is necessary to use the direct integration of Eq. (6.13). As a result, instead of Eq. (6.15), we get a different, simpler expression

$$J(0, u) = \frac{3}{4} \left[(1 + 2u + 2u^2) \ln \left(\frac{1 + u}{u} \right) - (1 + 2u) \right].$$ (6.39)

If we wish to be sure that the substitution of Eq. (6.39) into Eq. (6.3) leads to the well-known relation (4.9), from [195], for extinction close to the critical point, then it is enough to use the variable w in Eq. (6.39) (remember that $w = (2u)^{-1} = \left(\sqrt{2} k\xi \right)^2$ and consider that now $\gamma = 2\nu$ (see Eq. (4.35)).

According to the normalization condition in Eq. (6.10), $J(1, u) = 1$ holds as before. Moreover as $J(\zeta, u)$ weakly depends on ζ when $\zeta \approx 1$, then taking into account Eq. (6.37) we obtain

$$\frac{1}{l^*} \cong 1 - \frac{1}{w l^*},$$ (6.40)

where the precise equality corresponds to the Ornstein–Zernike approximation ($\eta = 0$), and the approximate one accounts for the fact that the Fisher index has to obey the inequality $\eta > 0$. As for the values of the quantity

$$\frac{\overline{\Gamma^*}}{l^*} = \int \Gamma^* dh^*,\qquad(6.41)$$

it is not possible to obtain a simple analytical expression, and therefore this integral has to be calculated numerically.

When $k\xi \ll 1$ for Eqs. (6.3) and (6.4), considering Eqs. (6.11), (6.12), and (6.33), and also due to the accepted normalization of Eq. (6.9), the approximate relation (6.42) holds

$$\frac{1}{l^*} = \frac{1}{l^*_{tr}} = 2^{1-(\eta/2)}(k\xi)^{2-\eta},\qquad(6.42)$$

and for Eq. (6.5), considering Eqs. (6.34), (6.11), (6.12), (4.3), and (4.26), due to the accepted normalization of Eq. (6.9), one can use Eq. (6.43)

$$\frac{\overline{\Gamma^*}}{l^*} = R\, 2^{2-(\eta/2)}(k\xi)^{1-\eta-x_{\overline{\eta}}}.\qquad(6.43)$$

6.2.1.3 Analysis of the Obtained Results [389]

The behavior of the nine quantities l^*, l^*_{tr}, Γ^*, $\overline{\Gamma^*}$, $(1-\overline{\mu})$, $(1/l^*)$, $(1/l^*_{tr})$, $(\overline{\Gamma^*}/l^*)$, and Γ^*_m is shown on Figs. 6.1–6.4. In semilogarithmic graphs, the dependence of all these quantities on $\lg \tau$ and the dimensionless parameter $k\xi$ in the variation range of the latter from 0.2 to 100 is presented. Recalculation is carried out using Eq. (6.44)

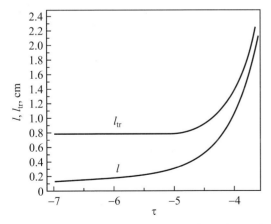

Figure 6.1 Dependence of the transport (l_{tr}) and normal (l) lengths of the free path of the photons on the degree of approach to the critical point of the binary mixture aniline–cyclohexane [389].

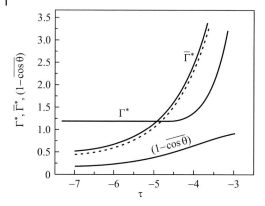

Figure 6.2 Dependences of Γ^*, $\overline{\Gamma}^*$ and $(1 - \overline{\cos\theta})$ on the degree of proximity to the critical point of the binary mixture aniline–cyclohexane [389].

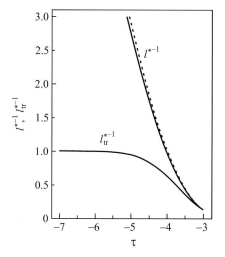

Figure 6.3 Dependences $h^* = l^{*-1}$ and $h^*(1 - \overline{\cos\theta}) = l_{tr}^{*-1}$ on the degree of approach to the critical point of the binary mixture aniline–cyclohexane [389].

$$\lg\left(k\xi\right) = \lg\left(k\xi_0\right) - \nu\lg\tau \tag{6.44}$$

taking into account Eq. (4.24).

In the computations, we used the parameters of aniline–cyclohexane mixtures $\xi_0 = 0.245$ nm, $n = 1.48$, provided that $\nu = 0.630$ [77, 102, 103], and $\lambda = 632.8$ nm. Because of the small variation in the values of ξ_0 and n for visible light scattering in different binary mixtures the range of variation of $\lg\left(k\xi_0\right)$ is not large: $-2.8 \leq \lg\left(k\xi_0\right) \leq -2.2$ [93]; aniline–cyclohexane having $\lg\left(k\xi_0\right) = -2.44$ virtually occupies the "central" position in it. To reduce Eq. (6.33) to a dimensionless form the characteristic length l_0 was used (see Eqs. (6.12) and (6.21)). By inserting

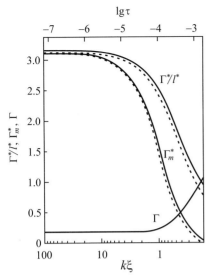

Figure 6.4 Dependence of (Γ^*/l^*), Γ_m^*, and Γ on the degree of approach to the critical point of the binary mixture aniline–cyclohexane [389].

the experimental data from [93] into Eq. (6.12) we obtain $l_0 = 0.71$ cm. In all the figures, the solid line corresponds to the results of calculation with $\eta = 0.031$ and $x_{\bar{\eta}} = 0.0635$. The dotted line gives the dependences corresponding to the Ornstein–Zernike approximation and therefore, to the universal scaling Kawasaki function (4.25), where $\eta = x_{\bar{\eta}} = 0$. The dotted line completely vanishes when for zero critical index values the result changes by less than 0.02.

Figure 6.1 demonstrates the difference of the transport free path length l_{tr}^* from the mean free path length l^* (see Eqs. (6.2) and (6.33)). When $k\xi \ll 1$ the scattering indicatrix is practically symmetrical relative to the plane, perpendicular to the incident beam. As a result $\overline{\cos\theta} \equiv \overline{\mu}$ close to zero and $l_{tr}^* \approx l^*$. It is clear from the figure that an increase in $k\xi$ leads to a continuous decrease in l^*. The transport length l_{tr}^* decreases significantly slowly and for large $k\xi$, starting from $\tau \sim 10^{-4}$, its value stabilizes close to unity. While l^* continues to decrease the ratio l_{tr}^*/l^* increases, reaching values for $k\xi = 2$ and $k\xi = 100$ equal to 2 and 7, respectively. Equation (6.2) shows that the reason for this can be found in the behavior of $\overline{\cos\theta}$.

Figure 6.2 illustrates the above assertion. On approaching the critical point (with a growth in $k\xi$) the scattering indicatrix is stretched extending forward and therefore $\overline{\Gamma}^*$ and $1 - \overline{\mu}$ decrease continuously. Let us emphasize the difference in the behavior of two spectral characteristics of single scattering, $\overline{\Gamma}^*$ and Γ^*. The first, which is averaged with respect to the scattering indicatrix, continuously decreases when $k\xi$ increases. The second, corresponding to scattering under a fixed angle θ, stays practically constant for large $k\xi$. The growth in the scattering indicatrix's asymmetry, as it approaches the critical point, can be estimated using the angle value $\theta_{\text{eff}} = \arccos(\overline{\cos\theta})$. Let us look at the results for some values of $k\xi$: $k\xi = 1, 10$

and $100 \Rightarrow \theta_{\text{eff}} = 72°, 43°, 32°$, respectively. When $k\xi \to \infty$ the minimal limiting value is $\theta_{\text{eff}} = 12°$ ($\eta = 0.031$).

Figure 6.3 deserves special attention as it demonstrates the qualitative difference in the behavior of two quantities, corresponding to the integrals (6.3) and (6.4). Formally, when $\eta > 0$ and $k\xi \to \infty$ both the integrals get close to finite maximum values, they are "saturated." In fact for the dimensionless extinction coefficient $h^* = l^{*-1}$ it is impossible to observe this effect. In order to reach even half the maximum of h^*, it is necessary that $k\xi = 10^9$. By contrast the quantity $h^*(1 - \overline{\cos \theta}) = l_{\text{tr}}^{*-1}$ already reaches half of its maximum value when $k\xi = 0.8$, and when $k\xi = 6$ its value is only 4% less than the maximum.

The main results, which correspond to Eqs. (6.36) and (6.41), are shown in Fig. 6.4. It is clear that $\overline{\Gamma}^*(l^*)^{-1}$ becomes saturated before $(l^*)^{-1}$: half of its maximum value corresponds to $k\xi = 0.4 - 0.5$, and when $k\xi = 4$ it is only 3% less than the maximum value. It should be noted that the difference between two versions of critical phenomena theories (solid and dashed lines, respectively) can be explained with the help of Eq. (6.43). When $k\xi > 10$ the value of Γ_m^* is stabilized. It is worth noting that the mean scattering multiplicity N continues to grow and when $k\xi$ changes from 10 to 100 N doubles its value. However, the simultaneous decrease in $\overline{\Gamma}^*$ (see Fig. 6.2) compensates exactly for this growth.

Figure 6.4 also shows that the limiting value Γ_m^* coincides for both versions of the theory and in accordance with Eqs. (6.27) and (6.35) we have $\Gamma_m^* \approx \pi R$ (in this graph it was assumed that $R = 1$). This means that by using the value Γ_m^* it is possible to estimate the difference of the universal amplitude R from unity. It should, however, be kept in mind that only the dimensionless values (Γ_m^*) behave in this way. The behavior of their dimensional analogues has already been discussed (see Eq. (6.31)). It should be remembered that for usual critical index values and typical values of $k\xi_0$, $\Gamma_{m,K}$ is practically twice as large as $\Gamma_{m,P}$.

The dependence of Γ_m^* on $k\xi$ (Fig. 6.4) describes the behavior of the scattered light spectrum provided that $N \gg 1$. In the experiment, we should be able to see a smooth change in scattering mode: far from the critical point, when $l \gg L$, single scattering is predominant and $\Gamma^* = (\Gamma/\Gamma_1)$, while in the immediate vicinity of the critical point where $l \ll L$, $\Gamma_m^* = 2(l_0^2/L^2)(\Gamma_m/\Gamma_1)$. It is clear that the left sides of these two formulae behave in opposite directions: Γ^* decreases when $k\xi$ increases (Fig. 6.2) while Γ_m^*, on the contrary, increases (Fig. 6.4). Such behavior was first obtained in [387, 389]. Till that time it has been assumed that when $T \Rightarrow T_c$ the half-width scattered light spectrum can only decrease.

In reality, a monotonic decrease in the spectral width can only be obtained when looking at scattering with a fixed multiplicity (single, double, etc.). Numerous papers on this topic are discussed in the review [350]. When the mean scattering multiplicity continuously increases the character of the changes in the spectrum half-width becomes fundamentally different. At first, when single scattering is dominant the behavior of the spectrum half-width corresponds to the well-known Kawasaki dependence. Then a smooth change in the scattering mode can be observed and Γ, which is now Γ_m, continues to grow and reaches its maximum value, as shown in Fig. 6.5 [387, 389]. This result appeared to completely contradict all the then existing conclusions of the theories of multiple scattering spectra close to the critical point. The next section is dedicated to investigate it experimentally.

6.3
Experiments Close to the Mixing Critical Point

Critical opalescence has been actively studied for many years (see, e.g. [26, 40, 117, 285, 287, 289, 292, 337–346, 364–382]), by methods of correlation spectroscopy including [93, 246, 247, 271, 275, 282, 319, 358, 359, 399–403, 489, 490]. However, as there was no theory for the spectral manifestation of developed critical opalescence, researchers always had to stay within the limits of low scattering multiplicity. After the development of a mathematical model of spectra of the diffusion mode of multiple scattering both on Brownian particles [386, 392] and on critical fluctuations [387–389] it became possible to go beyond these limitations and conduct experiments not only "far" from but also in the nearest vicinity of the critical point, in the range of continuously growing scattering multiplicity.

6.3.1
Experimental Setup

The setup used to study dynamic light scattering on model systems (see Section 5.3) was not designed to conduct experiments close to the critical point. That is the reason why many of its elements did not satisfy the necessary requirements for such investigations. First of all this is concerned with the accuracy of temperature measurement and maintenance. Moreover, the problem of PM afterpulses, which we have already discussed, required more fundamental modifications to solve it. As a result, the optoelectronic part of the setup was significantly changed. Let us briefly describe the essence of these changes [141].

In particular, a three-circuit thermostat with an electronic temperature control [493, 508] was created. A platinum resistance thermometer was placed inside the internal shell of the thermostat. In order to reduce thermal resistance, the free space cavity, where the thermometer was placed, was filled with oil. The thermostat system guaranteed a smooth temperature change of the sample within the range of 25–160 degrees, with an error not exceeding 0.3 mK. This made it possible to get closer to the critical point ($\tau \sim 10^{-5}$) and carry out measurements (see Fig. 6.5) in the nonhydrodynamic domain (right up to $q\xi \sim 5$).

We should not forget that the role of PM afterpulses sharply increases when weak signals are registered (see Section 5.2). In conditions of developed multiple scattering, independent of the reason for its appearance, the direct laser beam does not go through the medium but scatters completely in the whole volume. It should be emphasized again that this makes it possible to distinguish multiple scattering from scattering with a not very high multiplicity. Under these conditions, a significant part of the light is radiated on this part of the medium's surface,

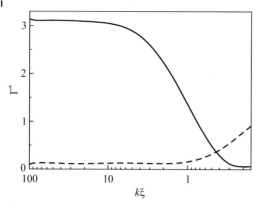

Figure 6.5 Comparison of the theoretical curves for the spectrum half-width of single (Kawasaki theory [307, 317] is the dashed line) and multiple (theory [387, 389] is the solid line) scattering.

surrounding an entering laser beam. From observation angles close to 60° (see below for more on the choice of observation angle) the scattered light intensity decreases sharply. This circumstance really complicates research as it aggravates even more the negative influence of PM afterpulses on the shape of the correlation function.

The usual method to cope with the effects of afterpulses, when performing the investigations close to the critical point, by significantly increasing the laser's power is, unfortunately, not acceptable. This is because in this area the system becomes exclusively "sensitive" to external actions. This last circumstance makes it necessary to take into account the possible heating of the medium by the laser beam [93, 326, 395, 494]. Such heating (for more see [395] and Section 6.4) is estimated, in various papers, to have the value of the order 1 mK/mW [93]. To eliminate this effect, the power of the laser beam in conditions of developed multiple scattering is usually decreased to a value of less than 1 mW.

To reduce the negative effect of afterpulses when investigating model systems, the data accumulated in the first few correlator channels were excluded from the correlation function approximation. However, this kind of data processing method can only be applied for to correlations times, larger than the time scales, typical for afterpulses (in our case 10 μs, see Fig. 5.3). When investigating critical opalescence the time correlation of scattered light reaches values significantly lower than 10 μs. Therefore, this method of accounting afterpulses is not acceptable for processing the correlation function.

Hence a special experimental method was used, namely cross correlation [495], which allows, in principle, to eliminate completely the contribution of afterpulses in the correlation function [418, 419, 496–498]. As can be seen in Fig. 6.6, when using cross correlation, the light scattered by the sample through an angle falls on the ray splitter (D, Fig. 6.6), thereby distributing the radiation between two photodetectors. As afterpulses arise in the two photodetectors independently no

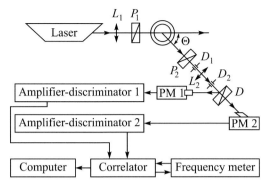

Figure 6.6 Diagramatic of experimental setup functioning with the cross-correlation method (the thermostat is not shown).

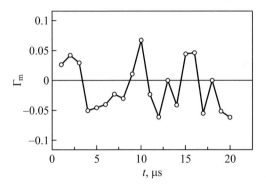

Figure 6.7 The correlation function of light, scattered on fixed scatterers using cross correlation [141].

correlation is seen between them and consequently, the correlation function is free from their influence. The correlation function from the fixed opaque screen in Fig. 6.7 demonstrates the efficiency of this method. It is clear that the cross-correlation method is really very effective in removing distortions caused by afterpulses (compared to Fig. 5.3).

In order to get a more effective accumulation of the correlation function, the intensity of scattered light falling on the photodetectors in two light beams should coincide. To create such conditions in practice is actually quite difficult. However, as shown in [498] this requirement is practically not critical. Nevertheless, in an attempt to create optimal measurement conditions the angle between the photodetectors was chosen close to 120°. It turned out that by so doing we managed to separate the scattered radiation intensity between both the PMs almost equally.

6.3.2
Choice of the Object of Research

By virtue of their general mechanism (anomalous fluctuation growth) critical phenomena are universal, and it would therefore appear that the choice of the object of research, within the limits of the same universality class, does not play any principal role. So, as pure liquids and mixtures belong to the same universality class ($d = 3, n = 1$) as the three-dimensional Ising model (see, e.g. [40]) it is possible to study critical opalescence either on pure liquids or on mixtures. However, this is not completely true. When the critical point is approached the system's susceptibility to external influences (gravitation, external fields, impurities, etc.) increases sharply. If we take into account the fact that they are less compressed, mixtures are less affected by the influence of gravitation than pure liquids, and also by selecting components with little density difference (see, e.g. [40]) this influence can be further decreased. Therefore, in this case, it would be better to use binary mixtures rather than pure liquids.

There are many different binary mixtures available which make them ideal subjects for studying critical opalescence. This makes it easy to choose the most suitable one for any kind of research. We needed a strongly opalescent system; so the main selection criteria were: (i) significant differences in the refraction indices Δn of the mixture components (scattering intensity $\sim (\Delta n)^2$) with (ii) similar values for density $\Delta \rho$ (weak appearance of the gravitational effect), and (iii) the mixture should have been well-studied previously.

The density of the components of the binary mixture aniline–cyclohexane [389, 391] is moderately different ($\Delta \rho = 0.24\,\text{g/cm}^3$) while there is a significant difference in the refractive indices ($\Delta n = 0.16$), so this mixture fully satisfies the formulated criteria. We can point out for comparison, that for weakly opalescent systems the analogous value of Δn is almost of an order of magnitude less. For example for the critical mixture 3-methylpentane–nitroethane $\Delta n = 0.0129$ and for nitroethane-*n*-hexane $\Delta n = 0.017$.

As for aniline–cyclohexane, it has probably been studied more than any other mixture. The analyses carried out on this system can be used to study the history of critical phenomena. Intensive research into this and other mixtures started in the mid-1950s (see, e.g. [117, 289] and references therein; [499] contains a good review of early work on critical opalescence). Since then binary mixtures have become frequently analyzed subjects to study the specifics of critical behavior similar to pure liquids.

6.3.3
Binary Mixture Aniline–Cyclohexane

One of the first experiments using the, at the time, newly created laser homodyne spectrometer was carried out by Cummins et al. [500] on the critical mixture aniline–cyclohexane, placed in a 3 mm diameter cylindrical cuvette. Alpert et al. [489] were one of the first to apply correlation spectroscopy, and the first to measure the width of the Rayleigh line close to the critical point in an aniline–cyclohexane mixture. This experiment confirmed the q^2-dependence of the half-width line close to the critical point (see Eq. (4.19)). However, extrapolation of the temperature dependence to T_c led the half-width line tend not to zero, but to a value equal to 25 Hz. In the discussion following this lecture[1] Alpert (working at Columbia Radiation Laboratory, Columbia University, New York), the author of [490], answering Marshall's (working at Atomic Energy Research Establishment, Theoretical Physics Division, Harwell, England) question concerning the reasons for this deviation replied that it would be eventually possible that they were not exactly at the critical point. To which Marshall remarked, "However, suppose that the scattering power of each molecule is slightly dependent on its neighboring molecules. Then the scattering power would depend on short-range order. You may then get a term which is in fact not divergent" ([5], p. 161). In our opinion this remark can be considered the first indication of the necessity of accounting for the background part for kinetic coefficients, and the data leading to this remark can be considered to represent its first experimental discovery. The first detailed theoretical and experimental papers on this topic appeared much later [318, 321, 325, 501, 502].

The dependence of the viscosity of this mixture on temperature close to the critical point was measured in [503] by using a capillary viscometer. It was known before this that a viscosity singularity exists [504]. However, it were the experimental data obtained in this paper that made it possible to make a quantitative assessment of the logarithmic character of this anomaly. It was a few years later that theoretical works [303, 307, 322] were performed where the viscosity singularity was explained and the corresponding critical index ($x_{\bar{\eta}}$) was calculated.

McIntyre and Wims, in their extremely thorough research [505] on static light scattering near the critical point of aniline–cyclohexane, and trying to get rid of multiple scattering, increased the sample's thickness up to 0.5 mm. When using a cuvette with thickness ~ 0.1 mm they found that the critical temperature changed by 0.5 °C. It seemed that this change was caused by the influence of boundaries, and/or by the action of surface forces. This example, to our mind, convincingly illustrates the pointlessness of trying to eliminate the influence of multiple scattering in this manner. On the other hand, the use of large cuvettes near the critical point has its own complications. Chu demonstrated that in cylindrical cuvettes with diameter of 10 mm or more, multiple scattering brings about significant distortions. However, the use of smaller diameters $\sim 2 - 3$ leads to significant complications due to light refraction and reflection [117].

An essential contribution to critical phenomena research, in general, and on the aniline–cyclohexane system, in particular, was made by Calmettes (see, e.g. [271–273, 314, 315]) and Beysens (see, e.g. [93, 95–97]). The dependence $\Gamma \propto q^3$

---

[1] As was already mentioned (see Chapter 1), the first large conference on critical behavior took place in 1965. Not only were the lectures published [5] but also transcripts of the numerous discussions between the participants. This part of the publication [5] was not less interesting than the lectures themselves. Unfortunately, this way of publishing conferences works did not become the norm.

(see Eq. (4.28)) was first found in [314]. It was after this that Kawasaki proposed his famous formula, Eq. (4.25). Investigation of the behavior of critical mixtures under shear flow was first carried out in [95] (see, also [96, 97]).

We can see that the mixture aniline–cyclohexane was really studied extremely thoroughly, using mainly optical methods. In studying this and other binary mixtures, researchers tried to eliminate multiple scattering in every possible way either by reducing the cuvette size or by completing the experiment far from the critical point, either choosing a weakly opalescent system or, finally, using the whole set of "security measures." In any case, according to existing information, the laser beam in the cuvette was always visible in all available experiments. This is a typical sign of insufficiently high scattering multiplicity. In the research discussed below [397, 398], opposite methods were undertaken: the increased cuvette dimensions, a strongly opalescent mixture, and significantly closer approach to the critical point.

It is well known that aniline is hygroscopic, oxidizes and acquires a yellow color in light, and over time turns completely black. Therefore, before putting together the critical composition, aniline is dried and repeatedly distilled in vacuum. If the cuvette with aniline is sealed off, no later than two hours after this procedure, it becomes completely transparent and colourless [271]. Only in this way the reproducibility of results can be guaranteed. As for cyclohexane, when spectral-grade, no extra cleaning is necessary. The prepared mixture has a concentration close to critical (43% aniline and 53% cyclohexane, by weight) and was dedusted by vacuum distillation in a glass cylindrical cuvette which was immediately sealed off. The cuvette's inner diameter was 50 mm, which guaranteed a noticeable increase in the scattering multiplicity even when the temperature was relatively far from critical. On the other hand, the cuvette's large optical thickness made it possible to carry out qualitative measurements under single scattering conditions very far from the critical point. This, as already mentioned above, is of primary importance for adequately accounting for the "background" part of critical scattering.

When constructing and preparing the thermostat, all possible measures were undertaken to reduce the temperature gradients to a minimum. A special test showed no temperature gradients along the cuvette's length in a stationary state within the limits of thermostatting accuracy. However, upon changing temperature they might appear. Temperature gradients, whatever their reason is, are always be accompanied by concentration gradients which grow as the critical point is approached. To eliminate their distortive influence from the experimental results, each measurement of Γ_m was carried out only after the system had been held at a given temperature for several hours. This measure made it possible to reach thermodynamic equilibrium in the system.

Another experimental difficulty when studying the critical point is the need to account for local heating of the laser beam in such a system. We managed to estimate the influence of laser radiation on parameters of the system close to the critical point, having analytically solved the medium heating problem under conditions of radiation power distribution due to multiple scattering (see Section 6.4). Moreover, upon transition to multiple scattering close to the critical point the light intensity, scattered under a certain angle, falls. This leads to an undesirable decrease in the signal/noise ratio and, therefore, to an increase in the signal accumulation time in the correlator. The accumulation time also increases as the power of the laser is decreased in order to minimize heating of the medium. As a result, when the power of the output beam does not exceed 1 mW, the measurement time of Γ_m at each temperature was about several hours. This made it necessary to ensure not only a high thermostatting accuracy but also the beam's stability over time. These conditions were satisfied in this experiment [397, 398].

6.3.4
Experimental Results

In Fig. 6.8 the full circles represent the measurement of Γ_m depending on $\lg \tau$ with a fixed observation angle $\theta = 60°$ [397, 398]. The observation angle θ was chosen as $60°$ for convenience of comparing the measurement results in single and multiple scattering modes. In fact, to describe the half-width spectrum for single scattering a wave vector of scattered radiation is used, while for multiple scattering incident (radiation) wave vector is employed. These vectors coincide with respect to their values at $60°$. The dashed line in the diagram represents the spectrum half-width of single scattered Γ at the same scattering angle $60°$. The calculation was made using the Kawasaki formula (4.23) [307].

Comparing the measurement results of Γ_m and the theoretical dependence of Γ allows us to single out two areas from the behavior of the half-width of scattered light. In the temperature range, corresponding to values of $\lg \tau$ in the range from

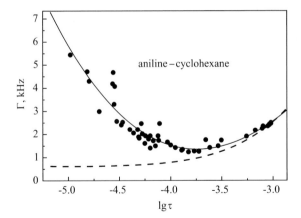

Figure 6.8 Temperature dependences of the half-width spectrum of multiple and single scattering modes: Full circles (•) show experimental data [397, 398], the dashed line the Kawasaki theory and the solid line is traced along the experimental points (see also Fig. 6.5 and the comment in the text to it).

−3 to −3.8, the half-width spectrum decreases monotonically, right up to the value $\lg \tau = -3.2$; the dependence fully corresponds to the theoretical curve for the single scattering mode. As the system approaches the critical point the scattering multiplicity grows. An increase in the scattering multiplicity leads to a monotonic growth in Γ_m, as demanded by the theory developed in [387, 389] and discussed in the previous section. Moreover, the quantitative comparison of the experimental and theoretical values of the "growth rate" of Γ_m on the virtually linear part of its sharp increase also gives a quite satisfactory agreement, in view of always high experimental error near to the critical point:

$$\left(\frac{\Delta \Gamma_m}{\Delta \lg \tau}\right)_{exp} = 6 \times 10^3 \, \text{Hz}, \quad \left(\frac{\Delta \Gamma_m}{\Delta \lg \tau}\right)_{theory} = 5 \times 10^3 \, \text{Hz}. \tag{6.45}$$

The above described experiment was the first to be carried out in the area of critical opalescence in the nearest vicinity to the critical point of a binary mixture [397, 398]. It showed that, in general, the behavior of the multiple scattering half-width spectrum corresponds quite well to that mathematical model which has been worked out on the basis of generalization of experimental information and theoretical ideas on multiple scattering on Brownian particles [141, 386, 390–394, 436–438, 440, 481] especially for this case [387, 389]. This experiment confirmed the basic conclusions of the theories [387, 389] that the multiple scattering spectrum half-width in this case, unlike single scattering or any other scattering with a fixed multiplicity, when approaching the critical point does not decrease but increase. At the same time "far" from the critical point, for single scattering, the experimental results [397, 398] agree very well with Kawasaki's well known and tested theory.

6.4
Heating of the "Critical" Medium by Probe Radiation

The problem posed in the title of this section is of great significance and interest when studying critical opalescence. This is because scaling equations like Eq. (1.4), by means of which the behavior of any physical properties of a system close to the critical point is described, demand exact knowledge of the distance from it with respect to temperature. Speaking about critical opalescence, we mean any research by optical methods as, even though the latest methods, owing to extremely weak interactions of light with matter, also related to invasive control methods, but when the properties of a substance strongly depend on temperature, the success of research, nevertheless, is impossible without properly taking into account medium heating by probe radiation. Estimates of the influence of such a heating of various systems near their critical points show that even for weakly opalescent binary mixtures (e.g., nitroethane–isooctane) the temperature increase may be as much as $\sim 1 \, \text{mK/mW}$ [93, 324].

In [494] the problem of heating the liquid crystal isotropic phase BMOAB (ButylMetOxyAzoxyBenzene) by probe radiation in the neighborhood of the nematic transition point was solved. Temperature distribution in the medium with absorption was obtained in the single scattering approximation. It is impossible, however, to draw any conclusions from these papers about the changes which can be expected when going over from single to developed multiple scattering. The problem of heating of weakly absorbing but strongly opalescent scattering media (e.g., binary mixture aniline–cyclohexane close to the demixing critical point) by laser radiation in such conditions was solved in [395].

Let a narrow laser beam fall onto a cuvette filled with a scattering medium. Penetrating into a cuvette, light dissipates in it, being partly absorbed which creates the volume distributed heat sources. When the scattering multiplicity tends to infinity, the passing light can be neglected and one has to take into account only the scattered light. It is necessary to find the temperature distribution in a cuvette provided that the boundary of the cuvette is kept at a constant temperature, T_0. The radiation diffusion equation, employed for the analysis of this problem, is

$$\Delta I = -\frac{3}{4\pi\ell_{tr}} P_0 \delta\left(\vec{r} - \vec{r}_0\right),$$ (6.46)

where I is, as usual, the light scattering intensity, $\vec{r}_0$ is the equivalent source location, and P_0 is the laser power. Following the sense and the determination of ℓ_{tr}, a direct laser light penetrates only to a depth of the order of magnitude of ℓ_{tr}. After that the knowledge concerning the initial direction of propagation is lost. Therefore, it is reasonable to choose the equivalent source position $(\vec{r}_0)$ at a distance ℓ_{tr}, counted off the boundary point down the beam direction. The boundary condition for Eq. (6.46) is

$$I(\vec{r})\big|_{r\in(\partial S + \ell_{diff})} = 0,$$ (6.47)

where $\partial S + \ell_{diff}$ is the surface, located at a distance ℓ_{diff} from the sample boundary in the direction of the outward surface normal.

If α is a radiation absorption coefficient then $\alpha I d\omega$ is a power of the heat sources in the unit volume, produced by photons with velocity directions distributed in the solid angle $d\omega$. Then $4\pi\alpha I$ is the full volume density of an energy source. In order to find the temperature distribution in the cuvette volume it is necessary to solve the thermal conductivity equation with a $4\pi\alpha I$ energy source

$$\Delta\Theta = \frac{4\pi}{c_p D_T}\alpha I,$$ (6.48)

where Θ is the temperature difference between the cuvette (T) and the thermostat (T_0). c_p is the heat capacity of the unit volume. The boundary condition for Eq. (6.48) is

$$\Theta\big|_{r\in\partial S} = 0.$$ (6.49)

Applying the Δ-operator to both parts of Eq. (6.48) and taking into account Eq. (6.46) we get

$$\Delta^2\Theta = \frac{3\alpha P_0}{c_p D_T \ell_{tr}}\delta(\vec{r} - \vec{r}_0) \equiv A\delta\left(\vec{r} - \vec{r}_0\right),$$ (6.50)

where

$$A = \frac{3\alpha P_0}{c_p D_T \ell_{tr}}. \tag{6.51}$$

The boundary conditions (6.47) and (6.49) must be mutually consistent. In order to solve Eqs. (6.46) and (6.48) with the boundary conditions (6.47) and (6.49), which would be the upper estimate for them, it is necessary to employ the boundary condition, Eq. (6.52), instead of Eq. (6.49)

$$\Theta|_{r\in(\partial S + \ell_{diff})} = 0. \tag{6.52}$$

Let us consider the Green function for the Helmholtz equation

$$\Delta I(\vec{r}, \vec{r}_0) - k^2 I(\vec{r}, \vec{r}_0) = \delta(\vec{r} - \vec{r}_0), \tag{6.53}$$

and write down the formal expression for I as

$$I(\vec{r}, \vec{r}_0) = \frac{1}{\Delta - k}\delta(\vec{r} - \vec{r}_0) = \Delta^{-1}\delta(\vec{r} - \vec{r}_0) + k^2\Delta^{-2}\delta(\vec{r} - \vec{r}_0) + O(k^4). \tag{6.54}$$

Now, it becomes evident that the multiplier at k^2 is an expression, which coincides up to a constant factor A with the solution of Eq. (6.50). In other words, to solve Eq. (6.50) it is sufficient to obtain the Green function for the Helmholtz equation with a precision up to k^2-order terms.

For a sphere with zeroth-boundary conditions on its surface this problem can be solved analytically for an arbitrary position of the radiation source. This solution is given in [395]. Here we will consider and analyze only the final result [141, 396]

$$\Theta(r, \varphi) = \frac{AR_1}{8\pi}\left[\left(1 + 2\rho\rho_0\cos\theta + \rho^2\rho_0^2\right)^{1/2} - \left(\rho^2 + 2\rho\rho_0\cos\theta + \rho_0^2\right)^{1/2}\right]$$
$$+ \frac{(1 - \rho^2)(1 - \rho_0^2)}{\rho^{3/2}\rho_0^{3/2}}\int_0^{(\rho\rho_0)^{1/2}}\frac{y^2}{1 + 2y^2\cos\theta + y^4}dy, \tag{6.55}$$

where

$$\rho = \frac{r}{R_1}, \quad \rho_0 = \frac{r_0}{R_1}, \quad R_1 = R + \ell_{diff}. \tag{6.56}$$

Here r_0 and r are the distances from the center of the sphere to the source and to the observation point, respectively, θ, as usual, is the scattering angle.

Let us derive an upper estimate for the expression (6.55)

$$\Theta(r, \varphi) \leq \frac{AR_1}{8\pi}\left[\left(1 + 2\rho\rho_0\cos\theta + \rho^2\rho_0^2\right)^{1/2} - \left(\rho^2 + 2\rho\rho_0\cos\theta + \rho_0^2\right)^{1/2}\right]. \tag{6.57}$$

Let us denote the expression in square brackets in Eq. (6.57) as H, then

$$\frac{\partial H}{\partial \theta} = \sin(\theta)\rho_0\rho\frac{(1 - \rho^2)(1 - \rho_0^2)}{F(\rho, \rho_0, \theta)}, \tag{6.58}$$

where

$$F(\rho, \rho_0, \theta) = (1 + 2\rho\rho_0 \cos\theta + \rho^2\rho_0^2)^{1/2}(\rho^2 + 2\rho\rho_0 \cos\theta + \rho_0^2)^{1/2}$$

$$\times \left[(1 + 2\rho\rho_0 \cos\theta + \rho^2\rho_0^2)^{1/2} + (\rho^2 + 2\rho\rho_0 \cos\theta + \rho_0^2)^{1/2}\right] \geq 0. \quad (6.59)$$

Thus, the derivative of H with respect to θ tends to zero when $\theta = 0$ and $\theta = \pi$ for any ρ, ρ_0, and also when $\rho = 1$ or $\rho_0 = 1$ for any θ. In the latter case $H = 0$ also holds, which makes it uninteresting. So, we have

$$H = \begin{cases} (1-\rho)(1-\rho_0) & \text{for} \quad \theta = 0, \\ |1-\rho_0\rho| - |1-\rho_0| & \text{for} \quad \theta = \pi. \end{cases} \quad (6.60)$$

It is clear that $H(\rho, \rho_0, \theta)$ reaches its maximum at $\rho_0 = \rho$ when $\theta = \pi$, i.e. at the source. Then the following estimate will be true:

$$H \leq (1-\rho_0^2)\big|_{\rho_0 \to 0} \cong 2(1-\rho_0), \quad (6.61)$$

$$\Theta(r, \theta) \leq \frac{AR_1}{8\pi} 2(1-\rho_0) = \frac{AR_1}{4\pi}\left(1 - \frac{r_0}{R_1}\right) = \frac{A}{4\pi}(R_1 - r_0)$$

$$= \frac{A}{4\pi}(\ell_{tr} + \ell_{diff}) = \frac{3\alpha P_0}{4\pi c_p D_T}\frac{5}{3} = \frac{5}{2}\frac{\Theta^*}{2\pi}, \quad (6.62)$$

where $\Theta^* = (\alpha P_0/c_p D_T)$ is a quantity with the dimensionality of temperature. For reasonable values of the parameters characterizing medium and source

$$\alpha = 0.1\,\mathrm{m}^{-1}, \quad P_0 = 2\,\mathrm{mW}, \quad c_p = 1.8\,\mathrm{J/(m^3\,K)}, \quad D_T = 1.5 \times 10^{-7}\,\mathrm{m^2/s},$$

$$(6.63)$$

we obtain $\Theta^* \cong 7 \times 10^{-4}\,\mathrm{K}$. Note that Θ^* does not depend on ℓ_{tr}. This feature can be qualitatively explained by the fact that the smaller is ℓ_{tr}, the longer are the photon's trajectories and consequently the absorption increases. On the other hand, the smaller is ℓ_{tr}, the smaller is the depth of the equivalent source position. These two tendencies compensate each other and as a result Θ^* does not depend on ℓ_{tr}.

Near the point of the falling beam, in the domain with a radius of the order of ℓ_{tr}, it should be noted that the diffusion theory for the radiation intensity will fail. Therefore, this domain demands special consideration [395]. Let a photon fall on the surface of a cuvette and then go to the boundary with a radius of ℓ_{tr} measured from the input point. Then the heat power which evolves in this volume is approximately equal to $\alpha\ell_{tr}P_0$. If we replace the distributed source by the point one which should be offset by l_{tr} from the boundary of the sphere, we get the following set of equations

$$\Delta\Theta = -\Theta^*\ell_{tr}\delta(\vec{r} - \vec{r}_0),$$

$$\Theta\big|_{|\vec{r}|=R} = 0, \quad (6.64)$$

where $R - r_0 = \ell_{\text{tr}}$. The solution to Eqs. (6.64) is

$$\Theta(r) = \frac{\Theta^* \ell_{\text{tr}}}{4\pi} \left[\frac{1}{(r^2 + 2rR\cos\theta + r_0^2)^{1/2}} - \frac{R}{(r_0^2 r^2 + 2rr_0 R^2 \cos\theta + R^4)^{1/2}} \right]_{r_0 \to R}$$

$$\cong \frac{\Theta^*}{4\pi} \frac{\ell_{\text{tr}}^2 (R^2 - r^2)}{(r^2 + 2rR\cos\theta + R^2)^{3/2}}.$$

(6.65)

The upper estimate of Eq. (6.65) will be

$$\Theta(r) = \frac{\Theta^*}{4\pi} \frac{\ell_{\text{tr}}^2 (R^2 - r^2)}{(r^2 + 2rR\cos\theta + R^2)^{3/2}} \leq \frac{\Theta^*}{2\pi} \frac{\ell_{\text{tr}}^2 (R - r)}{|\vec{r} - \vec{R}|^3},$$

(6.66)

where $\vec{R} = (0, 0, -R)$.

If we denote $\beta = \pi - \theta$ and use the obvious inequality $R - r\cos\beta \geq R - r$ we get

$$\Theta(r) \leq \frac{\Theta^*}{2\pi} \frac{\ell_{\text{tr}}^2 (R - r\cos\beta)}{|\vec{r} - \vec{R}|^3} = \frac{\Theta^*}{2\pi} \frac{\ell_{\text{tr}}^2 (\vec{r} - \vec{R})_z}{|\vec{r} - \vec{R}|^3}$$

$$= -\frac{\Theta^*}{2\pi} \frac{\ell_{\text{tr}}^2}{|\vec{r} - \vec{R}|^2} \cos(\vec{R}, \vec{r} - \vec{R}) = -\frac{\Theta^*}{2\pi} \ell_{\text{tr}}^2 f_z(\vec{r} - \vec{R}).$$

(6.67)

It should be noted that

$$f_z(\vec{r} - \vec{R}) = \frac{1}{|\vec{r} - \vec{R}|^2} \cos \vec{R}, \vec{r} - \vec{R})$$

(6.68)

can be considered as the z-component of a certain Coulomb force which acts on a unit charge at the point with the coordinates $(0, 0, -R)$ from the side of the volume element of a charged sphere with unit charge density. It is well known that in this case the full force acting from the side of a sphere does not change if this sphere is replaced by a point charge with magnitude $(4\pi/3)R^3$ located in the center of the sphere. After volume averaging, Eq. (6.67) can be finally written as [395, 396]

$$\overline{\Theta} = \frac{1}{(4/3)\pi R^3} \int\limits_{|\vec{r}| \leq R} \Theta(\vec{r}) dr \leq \frac{1}{(4/3)\pi R^3} \int\limits_{|\vec{r}| \leq R} \frac{\Theta^*}{2\pi} \ell_{\text{tr}}^2 f_z(\vec{r} - \vec{R}) dr$$

$$= \frac{\Theta^*}{2\pi} \left(\frac{\ell_{\text{tr}}}{R}\right)^2.$$

(6.69)

In the case of multiple scattering the quantity (ℓ_{tr}/R) is much less than unity. Then comparing Eqs. (6.69) and (6.62) one can see that the volume averaged contribution from this domain in the heating of scattering medium is negligible in comparison to the contribution of diffusively scattered light. So, we can make the theoretically very interesting and in practice extremely important conclusion: heating of matter, which is caused by multiple light scattering, when the free path transport length is small in comparison to the usual dimensions of the scattering volume $(\ell_{\text{tr}}/R) \ll 1$, does not depend on the scattering multiplicity.

This feature allows us to investigate the substance's properties using multiple scattering in the immediate vicinity of the critical point without continuously correcting the output power of the laser beam as it approaches the critical point. Furthermore, the calculation made here confirmed the estimates made by other authors [93, 324] in which the heating of weakly absorbing probe radiation systems near the critical point does not exceed $\sim 1\,\mathrm{mK}$ per mW of the output power of the laser beam falling into the investigated medium. It thereby becomes almost possible to correct the excess temperature differences which arise between the thermostat and the investigated medium close to the critical point due to heating caused by probe radiation, dependent on the single parameter, inserted in the medium power from the radiation source, without any particular consideration of the degree of approach to the critical point [395, 396].

7

Thermal Conductivity in the Vicinity of the Critical Point

7.1
Introduction

Among the transport coefficients of dense gases, thermal conductivity possesses one of the most clearly expressed anomalies near to the critical point. The viscosity anomaly is weaker and no experimental proof of the existence of the diffusion coefficient (self-diffusion) anomaly has yet been found [506]. As for the behavior of the mutual diffusion coefficient near the critical point, it was widely investigated by Matizen and colleagues using different methods (see, e.g., [507–509]). This analysis allowed them to propose an adequate description of diffusion processes in nonideal solutions.

The anomaly in the behavior of thermal conductivity was, it seems, first found experimentally in Guildner's 1958 paper [510] on CO_2 (quoted in [511]). It appeared that a significant increase in the thermal conductivity coefficient began just a few degrees apart from the critical point. At the distance to the critical point equal to $((T - T_c) \sim 1\,\mathrm{K})$, reached in this work, a 6–7-fold increase in the thermal conductivity coefficient (this effect is now called "critical enhancement") against a background of its smooth isothermal change on both sides far from the critical isochore was observed (a good impression on how such dependences look like can be gained from Fig. 7.1). A few years later, similar results were achieved, also on CO_2, in [512] and then on ammonia [513, 514]. Ammonia is famous in its own way. It was the first gas to be liquefied by simple cooling already in the second half of the 18th century [13].

As for the theoretical description of the behavior of thermal conductivity and diffusivity near to the critical point [515], it became clear, following the dramatic history concerning the high critical index values for thermal conductivity obtained by Benedek on SF_6 (see, e.g., [26, 246, 247, 319]) that to compare experimental data with theory it was necessary to take into account the background component of thermal conductivity as well as the critical ones [282, 320, 490, 501]. This point has already been discussed in Chapters 4 and 6. The papers [320, 501] show that the higher experimental critical index values for thermal conductivity compared to those predicted by theory ($\varphi = 0.74$ for CO_2 and Xe, and, in particular, for SF_6, for which $\varphi = 1.26$ instead of $\varphi_{\mathrm{theor}} \approx 0.6$), are caused by the ignored influence of the

Critical Behavior of Nonideal Systems. Dmitry Yu. Ivanov
Copyright © 2008 WILEY-VCH Verlag GmbH & Co. KGaA, Weinheim
ISBN: 978-3-527-40658-6

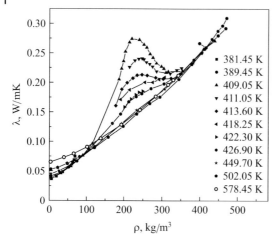

Figure 7.1 The parameters of temperature dependence of thermal conductivity of some substances.

regular part of the thermal conductivity coefficient. This situation can be corrected if a nondivergent background term is added to the thermal conductivity expression.

In [502], a method was worked out for estimating the background contribution to thermal conductivity and shear viscosity using experimental data obtained far from the critical point. As a result, the value of φ for CO_2 and Xe became close to the theoretical one, while for SF_6 further tests were required. After these new experiments, which included those of Benedek himself, it became clear ([319], pp. 346–350) that his previous data had not been fully precise, and actually the behavior of the thermal conductivity of SF_6 is not different from that of any other of the studied substances. So, from that moment, any remaining doubt concerning the necessity and effectiveness of separation of the thermal conductivity expression into two components (λ_b-background, regular, and $\Delta\lambda_c$-critical, divergent) was removed. We have to write, consequently,

$$\lambda(\rho, T) = \lambda_b(\rho, T) + \Delta\lambda_c(\rho, T). \tag{7.1}$$

Interestingly enough, the reverse effect also exists. With a sufficiently precise and sensitive experiment the nonmonotonic change of thermal conductivity in the vicinity of the critical isochore is found quite far from the critical point (see, e.g., [516–518]). For example, in [516], ∼2% critical enhancement was found near the critical isochore of argon ($T_c = 150.7$ K) even at room (!) temperature. Analysis of existing experimental data surprisingly shows that such behavior for "strong" singularities, like for c_p and K_T, is the rule rather than the exception (see also Section 2.2.9). It is well known that maximum values of c_p are experimentally found right up to a temperature twice the critical (see, e.g., [506, 519]). These features make the procedure of the background part separation quite a difficult task [517, 518].

The following arguments can be put forward to justify Eq. (7.1). Kadanoff and Martin [520] showed that kinetic coefficients can be represented as time integrals over the correlation function of the system (the so-called Kubo formulas [290], p. 539, [310], p. 532). Such expressions were first obtained by Green and Mori (see Section 3 of chapter 7 in [56]). The correlation function of the entropy density flux, by which the thermal conductivity coefficient near the critical point can be expressed, contains contributions from two fluctuating flows: one is the "classical" term, which arises due to *fast* fluctuations caused by molecular collisions, and the other is *slow*, which is related to the coupling of different modes, such as, interaction of entropy density flux with a transversal hydrodynamic velocity. As there is a very large difference in the timescale of both contributions, they can be considered as essentially independent. This feature is a physical basis of their separation and finally leads to Eq. (7.1).

The background part, λ_b, can, in principle, also be found theoretically. The thermal conductivity outside the critical area (λ_b) can be represented by the sum of two components $\lambda_b = \lambda_k + \lambda_v$, the first of which corresponds to a dilute gas, while the second is dominant when density is greater than the critical one (see, e.g., Section 3.10 in [521]). The first component (λ_k) can be calculated using previously mentioned kinetic theories [268, 269, 511], while theoretical methods developed for calculating the potential component (λ_v) are based on different models. In the case of a gas with a density greater than critical and in liquid state, the hard-sphere model (and its various modifications) can be taken as the most successful and general molecular theory for predicting transport coefficients of dense liquids. This model frequently gives a quite realistic description of molecular interactions (see, e.g., [521–524]).

Attempts are also being made to describe dynamic critical behavior by applying the modern renormalization group (RG) approach [26, 525–528]. A review of the results obtained using the RG-theory for critical dynamics in mixtures is presented, in particular, in [527]. This work also examines the specifics of temperature dependence of thermal conductivity and diffusion coefficients behavior, caused by temperature and concentration gradients, in the presence of long-range forces.

Kawasaki's mode-coupling (MC) theory [307] was further developed by Sengers et al. [529–531]. So in [529, 530], a solution of the MC-theory equations applied to critical fluctuation dynamics was proposed. This solution includes a crossover from singular behavior of the transport properties in the immediate vicinity of the critical point to the regular behavior far from this point. A modernized MC-theory which suggests the possibility of nonasymptotic critical behavior for transport properties is put forward in [531].

However, in such a complex area as the physics of nonequilibrium processes, experiment is always more complete than theory. And so for the treatment of the results of precise experiments, empirical and semiempirical methods [522, 532] are widely used, as well. In particular, in the theory and practice of dynamic critical phenomena for the dependence of background component of thermal conductivity (λ_b) on temperature and density it has long been supposed that one of its parts, related to ideal gas, can be represented by a function which depends only on temperature, and the other part relating to dense liquid is a function dependent only on density [501, 502, 533]

$$\lambda_b(\rho, T) = \lambda_0(T) + \tilde{\lambda}(\rho). \tag{7.2}$$

The next task is to investigate the peculiarities of the behavior of the kinetic coefficients of pure polar liquids in the neighborhood of the critical point based on the analysis of precise experimental data of thermal conductivity and scattered light for ammonia. According to its dipole moment (1.5 D, 1D $\equiv$ 1 debye = 3.33564 $\times$ 10^{-30} C m), ammonia is only slightly inferior to water (1.8 D) [511].

7.2
Thermal Conductivity of NH$_3$ Near to the Critical Point

Before we begin to employ Eqs. (7.1) and (7.2) in order to analyze the experimental results of [518], some general peculiarities in the methods of measuring thermal conductivity should be mentioned. Traditional methods of measuring transport properties are fundamentally connected with gradients of the corresponding quantities [534], which are introduced into the investigated system. It is, as a rule, difficult to adequately account for the various kinds of disturbances which thereby appear. The application of scattered light methods, on the contrary, generally gets rid of such inadequacies as there is no need to introduce any kind of macroscopic gradients. The spectrum of light, scattered on concentration and entropy fluctuations in the liquid, contains information concerning such transport coefficients as diffusion coefficients and thermal diffusivity, respectively (see, e.g., [319, 399–403, 535–538]). It thus becomes possible to divide naturally the area of research according to the applied methods. Near to the critical point, where the scattering intensity is large, the most effective methods for these goals are dynamic light scattering methods. Their use is based on the relation between spectrum broadening and the thermal diffusivity coefficient (see Section 4.3; it should be kept in mind, however, that in the immediate vicinity of the critical point the spectrum half-width has a complicated temperature dependence (see Chapter 6, Figs. 6.5 and 6.8). The application of traditional thermal conductivity measurement methods in this range, due to the above-mentioned reasons, is connected with the necessity to overcome significant experimental difficulties. Far from the critical point, on the contrary, light-scattering methods are not that effective, due to weak scattering. Therefore, by reasonably combining the two methods it is possible to study thermal conductivity behavior in a very wide neighborhood of the critical point. This possibility was demonstrated in the experiment on ammonia [518].

We have already mentioned that ammonia was the first gas which was successfully liquefied. In 1845, Faraday has already studied ammonia's vapor pressure and the liquid–vapor and liquid–solid equilibrium conditions [539]. The critical parameters were first defined (and quite precisely) in 1884 by Dewar [540]. The history of research on ammonia and data on its different thermophysical properties can be found in [541, 542]. All these investigations as well as the need to deepen and broaden knowledge of ammonia have taken on extra interest due to the Kyoto Protocol (1997), which stated, in particular, that all freons (perfluorocarbons) used generally as a refrigerants contribute to the "Greenhouse Effect" strengthening. Considering that a lot of freons were banned long ago, because of their destructive effect on the atmosphere's ozone layer by the Montreal Protocol (1986), it is quite possible that soon ammonia will be one of the few remaining refrigerants which can be used.

As for the study of the thermal conductivity of ammonia [518], initially information concerning a wide region of temperature and pressure change, the neighborhood of the critical point included, was quite small. In our opinion, the basic results can be found in [513, 514]. In these studies the measurements were carried out within the following parameter ranges, from −65 °C to 400 °C [513] and from 20 °C to 177 °C [514] for temperature and up to 39.5 MPa and 48 MPa with respect to pressure, respectively. In the area where critical effects could be already observed, the data from these papers differed significantly. In this connection, and considering an opportunity of special critical behavior of polar substances, as well (this question was discussed in detail in Chapter 2; see also [176–178]), a complex research of ammonia's dynamic properties in wide temperature (409–580 K) and pressure (1–80 MPa) ranges, the neighborhood of the critical point included, was performed in [518]. For this purpose, direct measurements of thermal conductivity and static light scattering were used. Measurements of the extinction coefficient with the help of light scattering, conducted near the critical point along the critical isochore, made it possible to obtain information about the critical amplitudes and indices Γ_0^+, γ, ξ_0 and ν, required for describing the critical behavior of ammonia's thermal conductivity (the purity of NH_3 (*Air Liquid*) was no less than 99.96%). As a result of using the data from light scattering together with that from direct measurements of the thermal conductivity it was possible to follow its changes along the critical isochore in the range 2×10^{-2} K $\leq (T - T_c) \leq 120$ K ($3 \times 10^{-5} \leq \tau \leq 0.18$).

7.2.1
Experimental Setup for Determining Thermal Conductivity

There are several well-known classical methods of measuring thermal conductivity [534]. The most popular ones are considered to be the slab, hot wire, and coaxial cylinders methods. Each of them has their own merits and disadvantages, but it is here not our task to critically analyze them. We shall only mention the fact that fairly qualitative results have been obtained in the most difficult area, the critical region, in different laboratories around the world by each of these three methods (see, e.g., [276, 516–518, 543–546]).

The Laboratory of Molecular Interaction and High Pressure (LIMHP)[1] of University "Paris-13," where our research was conducted [518], traditionally used the coaxial cylinder method and, it seems to us, could have almost made it perfect. Cylindrical geometry is particularly well suited for measuring at high and even

---

[1] The research in this chapter is based upon was carried out by the author during his year-long stay at LIMHP in 1978/79.

at extreme pressures. Works on determining the thermal conductivity of different gases [547–554] right up to 1 GPa [555], carried out at LIMHP, are unique (see, e.g., [556]). Also, the thermal conductivity data, obtained in this laboratory in the range of normal pressure, agree well with the results from analogous researches carried out at other laboratories using different methods [556]. It is also believed that the results obtained on the LIMHP setup, near the critical point, belong to the few reliable ones for pure liquids in this experimentally difficult area [556]. As details of this experimental setup can be found in [547, 550], we can concentrate here only on its essential features.

The cell for measuring thermal conductivity consists of two vertical coaxial silver cylinders, without a guard ring and with a heater placed along the axis of the internal cylinder. The gap between the cylinders is 0.26 mm. A stationary method is used. To avoid contact between the extremely electroconductive NH_3 and the electrical wires a two-gas system is used. The cell is hermetically sealed and connected to a bellow which is kept at room temperature.

The compressing gas (N_2) is introduced simultaneously outside the cell and the bellow so that the cell is kept free of high pressure effects. Nevertheless, to determine the possible corrections of the effect of pressure, the setup was calibrated by measuring the thermal conductivity of three noble gases, Ar (1 MPa), Ne (1 MPa), and He (10 MPa). Calibration was performed at a pressure higher than the atmospheric pressure to avoid correction due to accommodation effects. The heat losses by electrical leads and centering pings were determined by comparing the apparent and the most reliable values of the thermal conductivity of these fluids. The correction curve was assumed to be a function only of the thermal conductivity of the fluid under investigation. Corrections due to natural convection in thin fluid layers, the main source of errors when measuring thermal conductivity (see, e.g., [510, 557]), are generally minimized by performing experiments with small ΔT-values between the cylinders and reducing the vertical temperature gradient to a minimum. It is known that the heats, which are transferred due to thermal conductivity (Q_{tc}) and convection (Q_c), are proportional to powers of the temperature gradient in the first and second order, respectively. Thus, by varying the power supplied to the inner cylinder it is possible to confirm the absence of any noticeable convection. However, near the critical point such a method is not suitable because of significant changes of the average temperature which inevitably take place in the layer of the substance investigated. In this region the ratio of the respective heats was computed as

$$\frac{Q_c}{Q_{tc}} = C \operatorname{Ra} 2\pi r,$$ (7.3)

where C is the setup's constant, Ra the Rayleigh number, and r the mean radius of the fluid layer ($r \approx 1$ cm). The Rayleigh number is the product of the Grashof (Gr) and the Prandtl (Pr) numbers, i.e.,

$$\operatorname{Ra} = \operatorname{Gr} \operatorname{Pr}.$$ (7.4)

The Grashof number determines the convective heat transfer in the case of free convection when the motion is caused by density difference due to temperature nonuniformity near the heated matter. It can thereby be considered as a thermal modification of the Archimedes (Ar) criterion which takes into account the relation between buoyancy and viscous forces when there is initially a significant density difference

$$\text{Gr} = \frac{gl^3}{v^2}\beta_T \Delta T, \quad \text{Ar} = \frac{gl^3}{v^2}\frac{\Delta\rho}{\rho}, \tag{7.5}$$

where g is gravitational acceleration, l the typical system dimension, β_T the volume expansion coefficient, ΔT the temperature difference between the substances' surface and environment, v the kinematic viscosity, and ρ the density.

The Prandtl criterion, in turn, describes the ratio between the intensities of molecular transport of momentum and the heat transfer due to thermal conductivity

$$\text{Pr} = \frac{v}{D_T}. \tag{7.6}$$

The reason for introducing the Rayleigh number along with the Grashof criteria for consideration of the free-convective heat transfer

$$\text{Ra} = \frac{gl^3 \rho^2 c_p \beta_T}{\overline{\eta}\lambda}\Delta T \tag{7.7}$$

is related to the fact that, as numerical solutions to viscous heat-conducting medium equations show and experiment confirms, the dimensionless thermal conductivity coefficient (the well-known Nusselt criterion, Nu) is determined specifically by the product of the Grashof and Prandtl numbers, i.e., the Rayleigh criterion [558, 559]. In [518], maximum convection corrections, calculated according to Eqs. (7.3)–(7.6) and accounting for the properties of ammonia and the measuring cell, were found at the peak of thermal conductivity on the isotherm closest to the critical point ($T - T_c = 3.6\,\text{K}$) and did not exceed 2.5%.

Pressure was measured by two manometers calibrated for two ranges of pressure, 1–15 MPa and 0.1–150 MPa. The temperature difference between the cylinders, which for all temperature ranges was ~1 K, decreased to 0.15 K near to the critical point. In the described setup direct measurement of critical parameters was not provided for. However, it was possible to determine the values ρ_c and T_c by carrying out optical measurements (see below for more detail). The obtained values corresponded within the limits of error to those used in [542], i.e., $T_c = 405.4\,\text{K}$, $p_c = 11.33\,\text{MPa}$, and $\rho_c = 235.0\,\text{kg\,m}^{-3}$. This is the reason why these values were used for analyzing experimental data. It should also be noted, that data scattering of the NH₃ critical parameters in literature (see, e.g., the reviews [541, 542]) is generally not too large and in fact does not differ from those used in [518]. In our opinion, the main reason is that ammonia has not been so widely studied near the critical point as compared to other substances. Density was calculated using the equation of state for ammonia suggested in [542].

The accuracy of determination of temperature (pressure) was estimated with the help of two additional tests. Firstly, ΔT_{exp} was compared with ΔT_{calc}:

$$\Delta T_{\text{calc}} = \left(\frac{\partial T}{\partial p}\right)_{\rho_c}(p - p_c); \quad \text{for} \quad \text{NH}_3, \quad \left(\frac{\partial T}{\partial p}\right)_{\rho_c} = 4.81 \times 10^{-6}\,\text{Pa}^{-1}\,\text{K}, \tag{7.8}$$

where p is the pressure, measured at the peak of ammonia's thermal conductivity on two quasiisotherms closest to T_c. Secondly, the liquid–vapor phase transition curve was investigated. The fact that the system falls into a two phase region creates sharp changes to ΔT between the cylinders (see the curves corresponding to $T < T_c$ in Figs. 7.1 and 7.4), which made it possible to register the transition temperature very precisely. As a result the inaccuracy of the chosen value of T_c (equal to 405.4 K) was estimated as not exceeding $\pm 0.15 K$.

7.2.2
Experimental Results: Background Thermal Conductivity

The measured thermal conductivity data are shown in Fig. 7.1. In order to use them further it is necessary to adequately account for the background part of thermal conductivity, which, as has already been mentioned, consists of two components (see Eq. (7.2)). Thermal conductivity of rarefied gases, which is dependent only on temperature, can be represented in the following manner:

$$\lambda_0(T) = a + bT + cT^2 + dT^3 + eT^4. \tag{7.9}$$

where the coefficients having the values

$$a = 0.3589 \times 10^{-1}, \quad b = -0.1750 \times 10^{-3}, \quad c = 0.4551 \times 10^{-6}, \tag{7.10}$$

$$d = 0.1685 \times 10^{-9}, \quad e = -0.4828 \times 10^{-12}, \tag{7.11}$$

and the corresponding dimensionalities, were calculated on the basis of experimental data specially obtained in the temperature range 300–580 K [518]. This dependence in comparison to reference data is illustrated in Fig. 7.2.

The second background component of ammonia's thermal conductivity, which is dependent only on density, is also represented in a polynomial form as

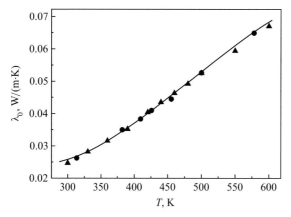

Figure 7.2 Thermal conductivity dependence of dilute NH_3 on temperature: • refer to data from [518], ▲ are reference data, and the solid curve refers to Eq. (7.9).

$$\widetilde{\lambda}(\rho) = \lambda_1\rho + \lambda_2\rho^2 + \lambda_3\rho^3 + \lambda_4\rho^4, \tag{7.12}$$

where the coefficients

$$\lambda_1 = 0.1621 \times 10^{-3}, \quad \lambda_2 = 0.1204 \times 10^{-5}, \tag{7.13}$$

$$\lambda_3 = 0.2314 \times 10^{-8}, \quad \lambda_4 = 0.3275 \times 10^{-11}, \tag{7.14}$$

having the corresponding dimensionalities, were obtained by least-squares fit from all thermal conductivity data on the isotherm corresponding to the temperature 577 K.

Although it was discovered [518] that in the case of ammonia, unlike Xe and CO$_2$, but like water, $\widetilde{\lambda}$ depends not only on density but also on temperature, it was nevertheless possible, when the temperature $T > 1.3T_c$ to consider $\widetilde{\lambda}$ as quasi-independent from temperature and to use Eq. (7.12). Therefore, as well as the data from the isotherm 577 K, the values of thermal conductivity on other isotherms were considered, on condition that $T > 1.3T_c$, and the density of ammonia exceeded the critical one at least twice. Figure 7.3 shows the dependence of the critical enhancement of the ammonia thermal conductivity on density at different temperatures after excluding the regular contribution determined by the described procedure. To compare the results with theory it is necessary to represent them as a dependence of the critical enhancement on the degree of approach to the critical point with respect to temperature along the critical isochore. In Figs. 7.4(a) and (b), the appropriate data are presented in ordinary and in log–log scale, respectively.

Here $\Delta\lambda(\rho_c) \equiv \Delta\lambda_c$ is reduced to a dimensionless form in the following manner:

$$\Delta\lambda_c^* \equiv \frac{\Delta\lambda_c}{\lambda^0}, \quad \lambda^0 \equiv \lambda_b(\rho_c, T_c) = \lambda_0(T_c) + \widetilde{\lambda}(\rho_c) = 0.126 \, \text{W m}^{-1} \, \text{K}^{-1}. \tag{7.15}$$

In Fig. 7.4(b), the presence of significant changes in the central part of the slope of the curve has provoked our attention. From the comparison of the data for ammonia [518] with the results of the detailed research carried out for argon and

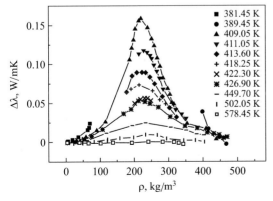

Figure 7.3 Critical enhancement of the thermal conductivity of ammonia as a function of density [518].

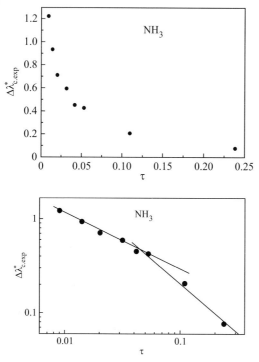

Figure 7.4 Critical enhancement of the thermal conductivity of ammonia along the critical isochore [518].

xenon, which was conducted almost simultaneously in the van der Waals laboratory (Amsterdam, Holland) [517] it was found [141] that a complete similarity in the behavior of all the three substances (Ar, Xe, and NH_3) near their critical points took place: the presence of a "kink" [517] in the curve (see Figs. 7.4(b) and 7.5). In [517], there is no graph like that shown in Fig. 7.5 for Xe but the authors confirmed that the behavior of Xe is analogous to that of Ar, although due to the greater data spread the kink in the curve is not expressed as clearly. Each of the separate "branches" in

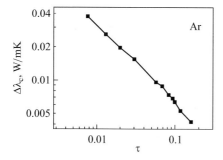

Figure 7.5 The critical enhancement of argon thermal conductivity along the critical isochore (using the data of [517]).

Table 7.1

Substance	φ_1	φ_2	Λ_1^*	Λ_2^*	τ_{kink}
Xe	0.62 ± 0.02	–	0.080 ± 0.007	–	–
Ar	0.64 ± 0.03	0.90 ± 0.07	0.059 ± 0.009	–	0.069
CO_2	0.63 ± 0.03	–	0.064 ± 0.007	–	–
NH_3	0.62 ± 0.03	1.12 ± 0.03	0.070 ± 0.006	0.015	0.053

the curve, given in Fig. 7.4(b) before (index "1" in Table 7.1) and after (index "2") the kink, were worked out mathematically [141] in accordance with the scaling law

$$\Delta\lambda_c^* = \Lambda^* \tau^{-\varphi}. \tag{7.16}$$

In Table 7.1, the obtained results are shown together with Trappeniers' data for Xe and Ar and also with the Sengers' data (1972) for CO_2 (quoted in [517]). The critical index φ_1 for NH_3 is obtained by averaging of values of the first six experimental points (see Fig. 7.4). In asymptotical vicinity to the critical point, this parameter decreases, as will be shown below, down to a value $\varphi_1 = 0.57$.

7.2.3
Extended Mode-Coupling (EMC) Theory

In his attempt to explain theoretically the appearance of a kink in the dependence of the thermal conductivity coefficient on the degree of vicinity to the critical point in temperature (Fig. 7.5), Trappeniers [517] employed the so-called extended mode-coupling theory (EMC). It is assumed in this approach [517] that "besides the contribution of the classical flux, there will appear in the critical region the effect of coupling between the streaming modes of the entropy-, the velocity- and the pressure-fluctuations, giving rise to anomalous contributions to the coefficient of heat conductivity." In general, the thermal conductivity coefficient can be presented as [517]

$$\lambda = \lambda_b + \Delta\lambda_I + \Delta\lambda_{II} + \Delta\lambda_{III}. \tag{7.17}$$

In this formula, $\Delta\lambda_I$ describes the effect of coupling of the entropy flux and the hydrodynamic velocity modes. Only the transversal velocity modes contribute to the singularity of the thermal conductivity. The longitudinal modes lead only to a nondivergent, minor effect which can be ignored ([51], pp. 616–621). The third term in Eq. (7.17), $\Delta\lambda_{II}$, is connected with the pressure–entropy correlation function. By virtue of the independence of the fluctuations of these quantities [27], this term is always equal to zero. The last term, $\Delta\lambda_{III}$, takes into account of the mode-coupling contributions of pressure and velocity fluctuations. This contribution is not only small but, in addition, vanishes at approaching the critical point. Thus, the anomalous behavior of the thermal conductivity coefficient near to the critical point is determined totally by the coupling of entropy and transversal velocity fluctuations. In the nearest vicinity of the critical point, the critical slowing down in the damping of thermal fluctuations is so large that the entropy correlation function may be taken assumed to be independent of time. Calculation of $\Delta\lambda_I$ under these conditions leads to the well-known result of Kawasaki's mode-coupling theory (MCT) [307] which has been discussed in Chapter 4 (see Eq. (4.23)).

As for the EMC-theory, its principal feature is the removal of the restriction that the entropy correlation function is time independent. By ignoring this restriction from the very beginning it was possible to obtain (B. W. Tiesinga, Thesis, University of Amsterdam, 1980; quoted according to [517]) a factor for $\Delta\lambda_I$ in the form of a correction function

$$F(a, b) = \frac{2}{\pi} \int_0^\infty \frac{d\omega}{(1+\omega)^2 \left[1 + a\left(1 + \omega^2\right) + b\left(1 + \omega^2\right)^{1/2}\right]}, \qquad (7.18)$$

where

$$a = \frac{\lambda_b}{c_p\overline{\eta}}, \quad b = \frac{\Delta\lambda}{\lambda}a. \qquad (7.19)$$

When τ is small ($\tau < 0.001$), this function is approximately equal to unity, i.e., the general expression goes over into the normal Kawasaki formula.

In [517], a numerical integration was carried out and tabulated values of $F(a, b)$ were given for argon at different temperatures, as well. It has been shown that, with the help of the new formula, corresponding to the EMC-theory, in contrast to the MC-theory, one can adequately describe the experimental data of argon in the whole measurement range of the thermal conductivity. Nevertheless, it should be noted that the smoothness of this correcting function (see Figs. 7.6(a) and (b)) cannot, in our opinion, solve the task for which this attempt was undertaken, that is to adequately account for the presence of a *kink* [517] in the dependence of the critical enhancement of the thermal conductivity on temperature. The very idea of extending the possibilities of the mode-coupling theory to describe experimental data on thermal conductivity over a large range of state parameters is, undoubtedly, promising (see also [529–531]).

As our main task in studying ammonia was to analyze the peculiarities of its critical behavior in the immediate vicinity of the critical point, an elementary correction function was chosen (see below for more). In the course of the new analysis [141] of ammonia data obtained in LIMHP [518], it became clear that the peculiarities found in [517] are, apparently, of universal character. It is possible that the author of [517] also assumed that this feature was universal since he tried to explain it from general positions. We can confirm this point of view by having a look at Figs. 7.7(a) and (b), which show in a log–log scale the dependence of the thermal conductivity critical enhancement on τ for different substances. The curves are plotted in [141] using tabular data presented in [554] (the study of thermal conductivity in the wide neighborhood of the critical point was started at LIMHP with ammonia [518] and then other substances were investigated [554]). It is clear that for all substances shown in Fig. 7.7 one can observe a similar behavior of the thermal conductivity critical enhancement on temperature. What is more, further

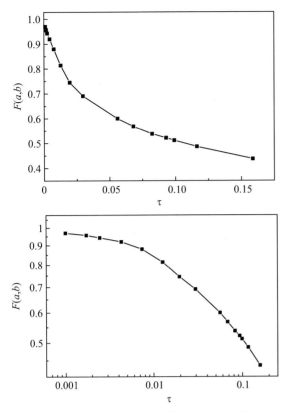

Figure 7.6 The correction function for argon in the form
$F(a, b) = c(1 - \tau)^d$ in normal (a) and in log–log (b) scale
[141]. The tabular data from [517] are employed.

analysis of this data showed [141] that they also all have two regions of scaling behavior with exponents φ_1 and φ_2 and values of the amplitude Λ_1^* very close to those shown in Table 7.1. However, the observed kink points do not show any clear dependence of their position with respect to the critical point temperature.

In [554], which was mainly dedicated to engineering approaches in describing the wide neighborhood of the critical point, these facts, for clear reasons, are not discussed. The analysis carried out in [141], apart from what has already been mentioned, shows that the kink on the dependence of thermal conductivity on temperature, independent on how explicitly it is expressed, is always present and for all studied substances $\varphi_1 = 0.55 \pm 0.07$ and $\varphi_2 = 0.9 \pm 0.1$.

What preliminary conclusions can be drawn from the conducted comparison of all the results on the critical behavior of thermal conductivity? First, it should be stressed that not only critical indices but also the amplitudes of such different substances, within the limits of error, coincide. This fact confirms, at least for thermal conductivity, the universality of critical phenomena dynamics which, as is known [51, 52], are not so predetermined and obvious.

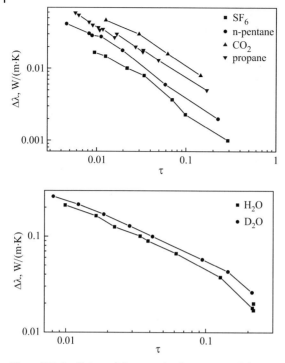

Figure 7.7 Coefficient of the critical enhancement of thermal conductivity of some substances along the critical isochore [141]. The plots are made using data from [554].

Another important conclusion is that there exist two different ranges of scaling behavior of the thermal conductivity coefficient near the critical point simultaneously and independently experimentally discovered for Ar and Xe [517] and NH_3 [518]. Moreover, after the analysis of the thermal conductivity behavior of a large number of different substances [141], presented in the extensive study [554], this experimental fact can now, seemingly, be considered as reliably confirmed. As a theoretical explanation of this behavior it could have probably been possible to use the EMC-theory [517] suggesting that the kink point corresponds to the moment when the condition of independence of the entropy correlation function from time is significantly violated. However, in any respects, this interesting and rather complex problem demands further theoretical research for it to be conclusively solved. As has already been mentioned, a detailed description of the thermal conductivity coefficient beyond the critical area is out of the scope of this book.

7.3
Static Light Scattering: The Extinction Coefficient

In the previous section, we outlined the procedure for direct ,determination and preliminary treatment of experimental data on the thermal conductivity of ammonia in the neighborhood of the critical point. To analyze these results in greater depth it was necessary to get additional information on some of its properties. This information was obtained through an optical experiment on static light scattering with the following determination of the extinction coefficient (h) [518].

The experimental determination of the extinction coefficient, fortunately, did not create any insurmountable technical difficulties and also made it possible to get sufficiently multifaceted information, especially near the critical point. Below we will show which specific information can be extracted from such an optical experiment and also discuss the results of new, more detailed analysis [141] of data, obtained in [518, 560]. These results shall be then used to discuss the behavior of ammonia's thermal conductivity in the neighborhood of the critical point.

The extinction coefficient represents itself the full scattering cross section and is determined according to Eq. (4.10), put forward in [195], to describe the data for SF_6 (it is commonly assumed that this equation first appeared in [195]). However, it was introduced there without any deduction; more general dependences (see Eqs. (6.15) and (6.39)) of which Eq. (4.10) is a special case, were, also first, developed in [389]. The basic formula used to determine it is Eq. (4.15). Therefore, it is necessary to know only the ratio of two light intensities, the incident on the studied system and that passing through it. The intensity ratios, unlike their absolute values, are fairly simple to measure with high precision, which is the reason for the relative simplicity of this method, as well as its hardware implementation.

7.3.1
The Experimental Setup for Light Scattering

The experiment with ammonia was carried out in a cylindrical steel cuvette. A thin laser beam ($\lambda = 632.8$ nm), less than 2 mm in diameter, was transmitted along its axis. Its power was held at the lowest level possible for the experiment, about 1 mW, so as not to allow uncontrolled heating of the studied substance due to the radiation passing through it (see Chapter 6 for more). Both ends of the cuvette had windows, sealed with Teflon, which guaranteed its hermeticity right up to a pressure of 15 MPa. The distance between the windows (3 cm) was chosen as a compromise between the demands of accurately determining the isothermal compressibility and the obtainable extent of measuring the temperature range without the appearance of multiple scattering. With this choice the measurements were restricted to the interval $(T - T_c) < 1$ K. The intensities of transmitted and incident light were measured by photodiodes, whose linearity was especially checked. Measures were also undertaken to avoid the negative effect of possible fluctuations of the laser's power on the results. The cell's volume (internal diameter 8 mm) was measured after being first filled with krypton at 25 °C and pressure 10 MPa and second with liquid ammonia at 18 °C and the same pressure. At these states of the phase diagram, the densities for both substances were calculated by their equation of state. Critical density was not measured and was taken equal to the literature value 235 kg/m^3 [542]. The temperature (405.4 K) at which the meniscus disappeared in the middle of the cuvette's height was taken as critical.

The cuvette was placed into a thermostat, a massive copper block. The thermostatting system guaranteed that the temperature was held with a precision no worse than 2×10^{-4} K. The measurement of the temperature deviation from the critical one was carried out using a quartz thermometer whose resolution and stability of 0.1 mK could only be realized with thermostabilization of the quartz generator itself.

7.3.2
Results and Analysis of the Optical Experiment

In Fig. 7.8, the experimental curve of the extinction coefficient (h) as τ dependence is shown in a log–log scale with the analytic continuation of its initial linear section ($h(0)$). The experiment was conducted on the critical isochore. As shown in Chapter 4, at the limit of small $k\xi$-values, far from the critical point, Eq. (4.10) goes over into Eq. (4.11), which makes it possible to determine the critical index of compressibility γ as the slope of the linear section of the presented dependence according to Eq. (7.20)

$$h^{(0)} = \frac{8}{3}\pi A K_T = \frac{8}{3}\pi A K_{0T} \tau^{-\gamma}. \tag{7.20}$$

The first ten experimental data, located far from the critical point on the linear section of this dependence, lead to the value

$$\gamma = 1.176 \pm 0.025. \tag{7.21}$$

Although this value is significantly lower than the theoretical value for the three-dimensional Ising model, $\gamma = 1.24$, the ideas put forward in this book's first two chapters show that it is more than reasonable for the measurement range $10^{-5} < \tau < 2 \times 10^{-3}$.

Equation (7.20) clearly shows that this dependence makes it possible, in principle, to determine, not only the critical index γ, but also the dimensionless amplitude $\Gamma_0^+ = K_{0T}p_c$. This parameter is just as important as γ. It is required for making

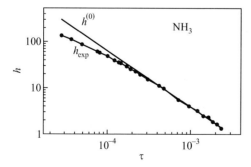

Figure 7.8 The extinction coefficient of ammonia depending on the vicinity to the critical point [141]. The curve is plotted using data from [560].

further calculations of the thermal conductivity coefficient as well as for analyzing the universal relation between amplitudes (see below).

7.3.3
Density Derivative of the Dielectric Constant

To compute the value of $\Gamma_0^+(K_{0T})$ using Eq. (7.20) it is first necessary, in accordance with Eq. (4.12), to calculate the value $\left(\rho\frac{\partial\varepsilon}{\partial\rho}\right)_T^2$, which is connected with local field effects. Theoretically, the dielectric constant and its density derivatives can be presented in a rather complex diagrammatic series [352] or expressed by many-particle correlation functions [341]. Unfortunately, it is impossible in both cases to obtain an analytical solution. Therefore, as a practical alternative, various model systems are usually employed for an interpretation. This is the way how different authors like Einstein (1910), Rocard (1928), Yvon (1937), and Vuks (1968) (see, for example, [93, 285, 288, 561]) accounted theoretically for the effect of the environment on the local dielectric constant. They all have based their own formulas on the connection of $\left(\rho\frac{\partial\varepsilon}{\partial\rho}\right)_T^2$ with the refractive index. The mentioned derivative can be presented in a general way as

$$\left(\rho\frac{\partial\varepsilon}{\partial\rho}\right)_T = S_n(n^2 - 1),\tag{7.22}$$

where S_n is the correction function for the refractive index, reflecting the connection between the scattered electric field and density fluctuations.

Einstein, using the Lorentz–Lorenz formula in its traditional form, obtained the result $S_n^E = ((n^2 + 2)/3)$ [287]. Rocard pointing out that the local field, in his opinion, is defined not only by the dielectric constant value in a small volume, but also by the dipole field of the entire volume of the liquid, treated the denominator $(n^2 + 1)$ in the Lorentz–Lorenz formula as constant [562, 563]. Rocard's result was $S_n^R = 1$. According to Vuks, the meaning of Rocard's criticism of Einstein's approach consists in the necessity to distinguish between the fluctuational and macroscopic values of the desired derivative [561]. By closely looking at this problem from the point of view of the theory of dielectrics, Vuks came to the following expression $S_n^V = (3n^2/(2n^2 + 1))$. By using the method of static correlation functions of the scattered field Yvon derived a formula [564], which, as a first approximation, did not differ from those Vuks later obtained from other considerations [561].

The practice of application of all these formulas shows that there is no unambiguous opinion on this subject. Thus, Calmettes in analyzing his data on the dependence $(\partial n/\partial T) = f(\tau)$ for aniline–cyclohexane preferred to use Rocard's formula [271–273], while Giglio and Vendramini found that their results for the extinction coefficient for the same mixture were described better by Vuks's than Rocard's expressions [565]. Beysens especially analyzed this problem studying light scattering in ten molecular liquids. As a result he came to the conclusion that all above theories correspond well to experiment except Einstein's [566] result. However, in a later paper Beysens and colleagues refined this result. This time they stated that Yvon's and Vuks dependences were more suitable for mixtures [567]. So, as we can see, every time we have to make a rather difficult selection.

However, close to the critical point this choice, strangely enough, becomes simpler. If we consider the van der Waals equation to be correct then the molecular refraction theory leads to the conclusion that the refractive index of any substance at the critical point should have a universal value, $n_c = 1.126$. Existing experimental data confirm this conclusion, pointing only to a small deviation from the theoretical value (see, for example, [288], p. 73). In particular, $n_c = 1.117$ is found for ammonia [560], which is actually not too far from the "classical" value. The difference is less than one percent. The difference between various S_n-values for ammonia with the use of this refractive index value is rather small, $(S_n^{max}/S_n^{min})^2 \cong 1.147$. The authors of [560] comparing the description of their results by Rocard's, Vuks's, and Lorentz's formulas came to the conclusion that the first of them gave a more realistic correction for ammonia. It should be mentioned that as the local field effect seems to substantially depend on the molecular structure of the liquid, then in different cases various methods of its account may be required. The existence of so many formulas for the solution of just one problem confirms that this is not such a simple and unambiguous task.

It turned out that for ammonia the difference in the determination of Γ_0^+, when using all the above formulas, is reduced to changes in the magnitude of the order of about $\pm 7\%$. In fact, the effect of choosing one or another value of S_n in the first approximation can be neglected in comparison to the role of errors in the estimation of other necessary parameters. In particular, as is shown in [141] (see also Section 2.3.3 in this book), critical indices and amplitudes are very strongly correlated. Reference to Eqs. (2.16) and (2.20) shows that an increase of γ by only 0.01 ($\sim$1%) results in a decrease in the value of Γ_0^+ by $\sim$7%. The real error in the determination of the index γ is, at best, a few percent, which leads to tens of percent error in the values of $\Gamma_0^+(K_{0T})$. Therefore, in preliminary calculations, as a correction to the local field effect in [141] the value determined by the Rocard formula was not used, but the average of the values given by all the formulas, and its refinement was put off until the establishment of the final results.

7.4
Determination of v and ξ_0 Using Light Scattering

One of the first experimental determinations of the correlation length with the help of light scattering was carried out by Skripov and Kolpakov near the critical point of SF_6 already in the "pre-laser" era in 1965 [82]. It was followed then by a variety of similar studies.

An important stage in the analysis was and is to adjust the function (see Eqs. (4.10) and (4.13))

$$F(w) = \frac{8}{3}\frac{h}{h^{(0)}} = \left[\frac{2w^2 + 2w + 1}{w^3}\ln(1 + 2w) - \frac{2(1 + w)}{w^2}\right], \tag{7.23}$$

which fortunately does not depend on K_{0T}, to experimental data. It can be seen from this formula that, although the function $F(w)$ uniquely depends on only one parameter $w = 2(k\xi)^2$, it is, in fact, in accordance with Eq. (4.24), a function of two other quantities, ξ_0 and v. This makes it clear that the experiment designed to determine the extinction coefficient, despite widespread statements of the contrary (see, for example [93]), does not make it possible to determine the value of these two quantities separately without additional assumptions. It makes it possible only to establish a certain their combination. Indeed, the value $F(w_i)$ can be constructed by a wide spectrum of correlated pairs of values ξ_{0i} and v_i (an analogous assertion relates to the combination Γ_0^+ and γ). Figure 7.9 demonstrates the correlation between the values ξ_0 and v found empirically [141].

It is considered to be useful to illustrate the discovered regularity between the amplitude and the critical index of the correlation radius using ammonia as an example [141]. Therefore, Figs. 7.10 show how does the quality of the approximation change if the correlated pairs of ξ_{0i} and v_i are chosen, i.e., satisfying the found dependence between them (Fig. 7.9), and if random pairs are chosen for the description of this kind of optical experiment.

It is easy to see that a good reproduction quality can be achieved when ξ_0 changes considerably, if, however, each time the corresponding value v (Figs. 7.10(b), (d)–(f)) is chosen. However, the quality of the approximation sharply worsens if the latter value is changed by a few percent (Figs. 7.10(a) and (d)). Figure 7.10(f) shows the deviation of the calculated values of ammonia's extinction coefficient found using the Levenberg–Marquardt approximation method program from those found experimentally. This unequivocally shows that the results of the computer computation should be treated with caution and, most importantly, with a large degree of healthy scepticism. Comparison of Figs. 7.10(e) and (f) once more demonstrates the fact that totally unrealistic values of the direct correlation length, ξ_0 (3 and 0.843 Å), correlated with the

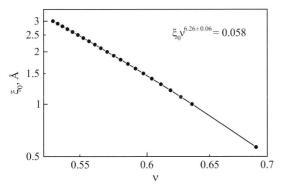

Figure 7.9 Correlation between the values ξ_0 and v for ammonia [141]. Curves are plotted according to data from [518, 560].

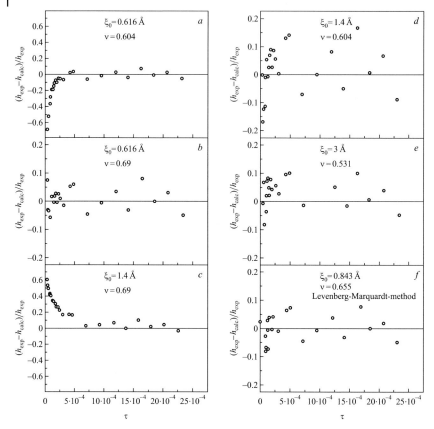

Figure 7.10 Comparison of calculated and experimental data for the extinction coefficient of ammonia with different combinations of values for ξ_0 and ν [141]. Curves are plotted using data from [518, 560].

corresponding values of the critical index, ν (0.531 and 0.655, respectively) give a wonderful coincidence of experimental and calculated data for the extinction coefficient.

The next two figures, Figs. 7.11 and 7.12, show the main results of investigation of the ammonia extinction coefficient, based on the experimental data from [518, 560]. It is quite natural then to ask why, out of the whole spectrum of possible pairs of values ξ_0 and ν, the pair

$$\xi_0 = (1.42 \pm 0.01)\,\text{Å}, \quad \nu = 0.606 \pm 0.002 \tag{7.24}$$

was chosen. As we have already mentioned, a light scattering experiment itself does not make it possible to determine both of these quantities separately without using additional experimental or theoretical information about one of them. Therefore, if only static light scattering experiments were carried out, it would be extremely difficult, if not impossible, to justify this choice. However, the

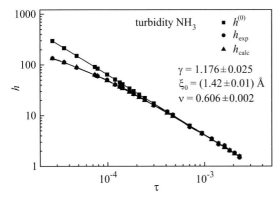

Figure 7.11 Experimental (h_{exp}) and calculated (h_{calc}) curves for the extinction coefficient of ammonia [141]. Plots are performed using data from [518, 560].

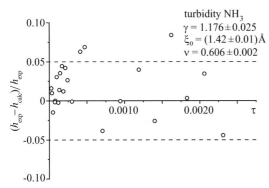

Figure 7.12 Deviation graph of calculated and experimental data for the extinction coefficient of ammonia with an optimal combination of values for $ξ_0$ and v [141]. Plots are drawn using data from [518, 560].

results of a comprehensive analysis of ammonia's thermal conductivity, along with the analysis of universal relations between critical amplitudes and indices, which will be discussed later (see also Chapter 2), makes this choice still difficult but not hopeless. As with all possible choices, this one was clearly the result of a certain compromise. It will be discussed further when Eq. (7.27) is analyzed.

Another obvious but more simple question could be formulated as "what is an error of a hundredth or even a thousandth of a fraction if it would be possible to choose other values which are just as good but significantly different from the suggested ones?" The answer is that in this case the shown error, in fact, only determines the boundaries for the values of $ξ_0$ and v, outside of which the quality of the approximation is noticeably worse.

7.5
Critical Dynamics: Comparison of Theory and Experiment

The dynamics of critical phenomena, whose fundamentals were set out in Chapter 4 (see Eqs. (4.17)–4.29)) makes it possible to express the coefficient of thermal conductivity in the neighborhood of the critical point as

$$\Delta\lambda_c = D_T \rho c_p^c = R \frac{k_B T}{6\pi\bar{\eta}\xi} \rho c_p^c, \tag{7.25}$$

where the universal dynamic amplitude R has different values depending on the calculation method. In the MC-theory $R = 1$ holds, and the RG-theory suggests that $R = 1.2$. This question was also discussed in Chapter 4 (see Section 4.4).

To compare Eq. (7.25) with experiment, the critical part of the isobaric heat capacity c_p^c is usually replaced by $(c_p - c_V)$. There are several reasons for this (see, e.g., [40]). Firstly, the value in Eq. (7.25) on the left side represents the difference of the total thermal conductivity coefficient (λ) and its regular part (λ_b), and the latter, as known [40, 501], is proportional to ρc_V. Secondly, unlike for c_p, for $c_p - c_V$ there exists a precise thermodynamic expression in terms of the experimentally measurable quantities. Finally, in the critical region, c_p is tens of times greater than c_V, so in any case $c_p \approx (c_p - c_V)$.

When describing thermal conductivity data far from the critical point it is c_p, the value of which can be calculated directly according to the suitable equation of state, and which should be used. By applying standard thermodynamic methods it can be easily s that

$$c_p^c \cong c_p - c_V = \frac{T}{\rho} \left(\frac{\partial p}{\partial T}\right)_{\rho_c}^2 K_T. \tag{7.26}$$

So when using dimensionless units and neglecting the viscosity anomaly, Eq. (7.25) can be rewritten as

$$\Delta\lambda_c^* = R \frac{1}{\lambda_b} \frac{k_B T_c^2 (1+\tau)^2}{6\pi\bar{\eta}_b \xi_0 \tau^{-\nu} (1+a_\xi \tau^\Delta)} \left(\frac{\partial p}{\partial T}\right)_{\rho_c}^2 K_{0T} \tau^{-\gamma} (1+a_\chi \tau^\Delta), \tag{7.27}$$

where

$$\left(\frac{\partial p}{\partial T}\right)_{\rho_c} = [2.18 - 0.12\exp(-17.8\tau)] \times 10^5 \text{ Pa K}^{-1} \tag{7.28}$$

according to the equation of state for ammonia [542];

$$\Delta = 0.5, \quad \frac{a_\xi}{a_\chi} = 0.7, \quad a_\chi = 1 \tag{7.29}$$

[50]; the regular part of viscosity is represented by

$$\bar{\eta}_b = (2.60 + 1.6\tau) \times 10^{-5} \text{ N s m}^{-2} \tag{7.30}$$

[518]. The critical index of compressibility is $\gamma = 1.176 \pm 0.025$ (Eq. (7.21)). The critical temperature was initially treated as a free parameter but further analysis showed that the optimal result can be obtained using the value $T_c = 405.4$ K, determined experimentally. The choice of the other quantities was performed in the following way.

7.5.1
Universal Dynamic Amplitude R

Based on the parameter "A" (see Eq. (4.12)) obtained from the experiment for determining the extinction coefficient and on the value chosen for $\left(\rho \frac{\partial \varepsilon}{\partial \rho} \right)_T^2$, the calculated amplitude value K_{0T} turned out to be equal to $K_{0T} = 6.389 \times 10^{-9}$ Pa^{-1} [141]. This value was further considered as only a first approximation. It is impossible to judge the degree of "correctness" of this amplitude value as it is different for all substances. The dimensionless amplitude of isothermal compressibility Γ_0^+ is correlated with the critical index γ. By using the found value of K_{0T} and critical pressure $p_c = 11.33$ MPa we get $\Gamma_0^+ = 0.072$ for ammonia. This result appears quite reasonable when $\gamma = 1.176$ and it agrees well with similar amplitude – index combinations for other polar and nonpolar substances (see Fig. 2.28). The universal dynamic amplitude R was initially taken as equal to unity. Then this value has been adjusted by means of the amplitude–index correlation (2.16). It is evident from Fig. 2.28, corresponding to this dependence, that, keeping in mind a preset value of critical index $\gamma = 1.176$, the compressibility critical amplitude should not leave a range $\Gamma_0^+ = 0.073 \pm 0.002$. Then corresponding calculation shows that the corrected universal dynamic amplitude appears equal $R = 0.99 \pm 0.03$. It should once more be mentioned that errors in the values of $\xi, \nu, K_0 T$, and R only define the corridor outside of which the quality of the approximation is clearly affected. In particular, for example, it was not confirmed that the universal dynamic amplitude was established and really equal to $R = 0.99 \pm 0.03$ (this would have been an extremely important result). All we are saying is that it *might* be equal to this only in combination with the noted values of other quantities and then even a one percent change clearly affects the quality of the approximation.

The only thing that what has just been said does not relate to is the index γ (see (7.15)), whose error is calculated according to all the rules and should therefore be taken in a general sense. To illustrate this Figs. 7.13(a), (b), and 7.14 are used to compare the approximation results when the index γ changes by as little as 0.004 with an overall error of 0.025, and demonstrating the *sensitivity* of this procedure to even the smallest parameter variation.

7.5.2
Thermal Conductivity Critical Index, φ

This critical index was determined from a direct measurement of the thermal conductivity coefficient via the slope of the tangent plotted along several points, closest to the critical, to the line of its temperature dependence (Fig. 7.13(a)). As this method is not very accurate due to the small number of points calculated, the obtained value $\varphi = 0.60$ can also be considered as only a first approximation.

As a result of performing a new estimate of experimental data for the critical enhancement of thermal conductivity involving universal relations between amplitudes, it turned out that $\nu = 0.606 \pm 0.002$ is the optimal value for the

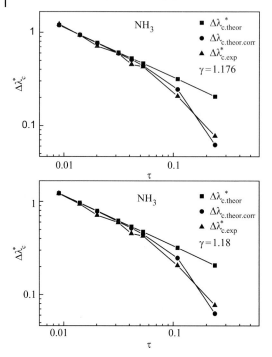

Figure 7.13 Coefficient of critical enhancement of the thermal conductivity in the wide neighborhood of the ammonia critical point.

critical index of the correlation radius. Afterward, optimal values for other desired properties were practically uniquely determined as $\xi_0 = 1.42 \pm 0.01$ Å and $\varphi = \gamma - \nu = 0.570 \pm 0.002$ instead of $\varphi = 0.60$, which was determined as the first approximation.

It then became clear that from all the possible values of the product $R K_{0T}$ the best is the greatest of them, that corresponds, like in [518], to Rocard's representation for $\left(\rho \frac{\partial \varepsilon}{\partial \rho} \right)_T^2$.

7.5.3
Checking the Feasibility of the Universal Relations Between the Critical Amplitudes for Ammonia

By substituting the necessary parameters into Eq. (2.14) we obtain $R_\xi^* = 0.719$ for ammonia, which agrees well with theoretical value: $R_\xi^* = 0.668$. As to the universal combination R_χ^+, for calculation of its value for ammonia using Eq. (2.13), let us at first define possible amplitude D_0-value, starting from correlation dependence $D_0 - \delta$ (Eq. (2.17)). It is evident from Fig. 2.29, corresponding to this dependence, that, keeping in mind a preset value of critical index $\delta = 4.30$, the critical amplitude D_0 should not leave a range $D_0 = 1.70 \pm 0.01$. Corresponding calculation shows

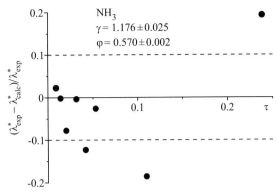

Figure 7.14 Deviation graph of calculated and experimental data for the thermal conductivity coefficient when an optimal combination of values for γ and φ is chosen [141].

that then R_χ^+ turned out to be equal to $R_\chi^+ = 1.53 \pm 0.09$. This result agrees well within the limits of error with the value $R_\chi^+ = 1.44$ obtained from the analysis of the big array of experimental data (see Section 2.3.3), and with its RG-value $R_\chi^+ = 1.60$ [76]. On the other hand, if we take both theoretical values

$$R_\chi^+ = \begin{cases} 1.60 & \text{RG-method} \\ 1.75 & \text{HTS-method} \end{cases} \tag{7.31}$$

we get, with the help of Eq. (2.13), $D = 1.77$ and $D = 1.62$ for the amplitude of the critical isotherm, respectively.

7.5.4
Thermal Conductivity of Ammonia in the Wide Neighborhood of the Critical Point

The theoretical description of the excess critical thermal conductivity, which is based on the ideas of the mode-coupling theory, the RG approach, or other model representations, is justified only in the asymptotic proximity to the critical point. in order to widen the range of applicability of these theories, attempts have been made to modify them. For example, an attempt was made to develop the EMC-theory [517], using the experimental data for thermal conductivity of argon. More often, however, various empirical or semiempirical correcting functions were introduced [568]. As the correcting function, obtained with the EMC-theory, turned out to be nonuniversal and as our task does not include, in principle, research into the wide neighborhood of the critical point, a second way was chosen. Thus, like in [518] a correcting function was used of the form

$$F_{\text{corr}} = \exp(-f\tau^2), \tag{7.32}$$

where f is an adjustable parameter. Then $\Delta\lambda_{c.\,\text{exp}}^*$ can be represented as

$$\Delta\lambda_{c.\text{exp}}^* = \Delta\lambda_{c.\text{theor}}^* F_{\text{corr}}. \tag{7.33}$$

From Eq. (7.33) it is clear that the correcting function, chosen in this way, does not affect the process of comparing experimental data on thermal conductivity with theories near to the critical point. It also makes it possible to broaden the experiment's description, with the help of the theories, to the region far from the critical point (a better approximation for ammonia was achieved when $f = 21$ [141], instead of $f = 36$ [518]).

7.6
Conclusion

The complex research carried out on ammonia [141, 518, 560] in the wide neighborhood of the critical point made it possible to compare the obtained results with existing theoretical concepts and to obtain a fuller picture of the critical behavior of the kinetic coefficients of polar liquids. It turned out that for ammonia

- In the asymptotic region close to the critical point the critical index of thermal conductivity $\varphi = 0.570 \pm 0.002$ agrees well with modern theoretical results.

- The value of the critical index of isothermal compressibility for polar ammonia $\gamma = 1.176 \pm 0.025$, found using optical methods in the temperature range $5 \cdot 10^{-4} < \tau < 2 \times 10^{-3}$, being significantly lower than the theoretical value, agrees completely with data on nonpolar SF_6 (see Eq. (2.12)). This fact is totally in agreement with the ideas which were put forward in the first two chapters of this book about the possible change of this critical index on approaching the critical point due to the effect of various disturbing factors. In the case of ammonia these factors could be both gravitation and Coulomb forces caused by the high value of its dipole moment.

- The value of the correlation radius critical index, ν, obtained as a result of trying to satisfy most of the relations where this index enters, at high quality of approximation of experimental data on heat conductivity led to the value $\nu = 0.606 \pm 0.002$. This value is also lower than the theoretical one, in our opinion, for the same reasons as for the index γ. The values of both of these indices correlate well with each other and thereby, give the rare possibility to calculate the Fisher critical index (see Eq. (4.35)) via the relation $\eta = 2 - (\gamma/\nu) \cong 0.059$.

  It is well known that there is an uncertainty in experimentally defining this index [332, 333] (for more see Chapter 4) so the obtained value should, probably, be considered as quite corresponding to theory. It should be remembered that in three dimensions, using all calculation methods, its value is located between $0.031 < \eta \leq 0.041$ (see, e.g., [40]).

- In the process of analyzing the results of the optical experiment in determining the extinction coefficient, a correlation between ξ_0 and ν was established. For ammonia the found empirical connection between them (see Fig. 7.9) can be expressed as a power law dependence [141]

$$\xi_0^* \nu^{6.26\pm0.06} = 7.34 \times 10^{-3}, \tag{7.34}$$

where

$$\xi_0^* = \frac{\xi_0}{a}, \quad a \equiv \left(\frac{k_B T_c}{p_c}\right)^{1/3} = 0.79\,\text{nm} \tag{7.35}$$

It should be mentioned, that it was the dependence 7.34 that marked the beginning of our research on correlations between various critical indices and amplitudes (for more see Section 2.3.3 and [226, 237])

- K_{0T} and Γ_0^+ turned out to be equal for ammonia $(6.39 \pm 0.18) \times 10^{-9}\,\text{Pa}^{-1}$ and 0.072, respectively. The latter value agrees well with the analogous value for SF$_6$, as well as with the general features which connect Γ_0^+ and γ (see Section 2.3.3).

- Within the limits of accuracy of the conducted research on the dynamic critical behavior for pure liquids using ammonia as an example, it was also established that a better agreement with light scattering experimental data was achieved if for $\left(\rho\frac{\partial\varepsilon}{\partial\rho}\right)_T^2$ Rocard's formula was used. However, it seems that for mixtures the Yvon–Vuks approximation is more precise [566].

- As for the universal dynamic amplitude its value was equal to $R = 0.99 \pm 0.03$, which agrees better with the Kawasaki–Ferell–Paladin–Peliti representation and the experimental value that Beysens and colleagues finally arrived at, i.e., $R = 1.06 \pm 0.06$ [324, 330], than with the RG calculation within the scope of the so-called H-model, where $R = 1.2$ [322] (see also Chapter 4).

- The result that we obtained for the index γ, $\gamma = 1.176$, taken together with the critical index $\delta = 4.3$ from [542], in accordance with Eq. (2.4) leads to the value of the coexistence curve critical index $\beta = 0.356$ for ammonia. Latter result agrees well with this index's experimental value obtained for ammonia in nearly the same temperature range $((T - T_c) < 0.5\,\text{K})$: $\beta = 0.35$ [569], as well as with the value of β for SF$_6$, obtained in the (pVT)-experiment [87, 89, 91, 141] (see Eq. (2.11)). In this case, the amplitude of the coexistence curve, using data from the same source [569], $B_0 = 2.15$, despite an overestimation of the value, nevertheless still agrees quite well with the data for SF$_6$ (see Eqs. (2.12) and [141]) and with the correlation dependence which connects β and B_0 (see Section 2.3.3).

Overall, the results of the analysis allow us, with a large degree of certainty, to suggest that the dynamic critical behavior of polar and nonpolar pure liquids in states with a significant influence of various disturbing factors has some general common features. In particular, all the universal relations, which exist between critical indices and amplitudes, are satisfied, despite the presence of individual peculiarities in the deviations of experimentally determined critical indices from their theoretical values.

Appendix A

Some Applications of the Photon Correlation Technique

A.1
Diffusing-Wave Spectroscopy

An important stage in the development of dynamic multiple scattering in the photon correlation technique is diffusing-wave spectroscopy. Its first real occurrence was in 1987 in [441], where backscattering from a substance with high optical density in conditions of flat slab geometry was studied. In their research, Maret and Wolf looked at the Brownian motion of the scatterers and applied a diffusion approach just as had been done earlier [386]. But if the paper [386] passed almost unnoticed the appearance of the paper [441] led to an "avalanche" of papers which is still continuing (see, e.g., [441–463]).

One of these papers [442] also has given the name of this modification of the dynamic scattering method, i.e., *diffusing-wave spectroscopy*. This technique makes it possible to successfully study such media as dense colloidal suspensions and foams, emulsions and granulated media, biological, medical and food products, etc., guaranteeing in each case a sufficient time and spatial resolution [452, 460]. The geometry of diffusing-wave spectroscopy presumes illuminating the flat layer with a light wave (Figs. A.1 and A.2). The advantage of organizing an optical experiment in this way is that the diffusion equation for photons can, in this case, unlike for cylindrical geometry, be solved precisely [446, 448, 452, 463]. Moreover, in this case the whole autocorrelation function, and not just its initial slope (the first cumulant), is registered. One of its shortcomings is the need to apply quite a powerful laser because of the wide light beams (this makes this technique almost useless for studying critical phenomena) and also the difficulty of changing the mean scattering multiplicity due to the constant thickness of the sample.

Unlike slab geometry, cylindrical geometry does not require a powerful laser. It is easy to change the scattering multiplicity by simply changing the observation angle. Although the analytical solutions of the theory obtained using cylindrical geometry cannot be expressed via elementary functions, there is always, in principle, the possibility to carry out the experiment in geometrically similar conditions to exclude unknown factors, as shown in Chapter 5. The merits of slab and

Critical Behavior of Nonideal Systems. Dmitry Yu. Ivanov
Copyright © 2008 WILEY-VCH Verlag GmbH & Co. KGaA, Weinheim
ISBN: 978-3-527-40658-6

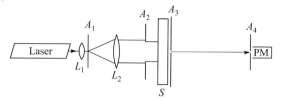

Figure A.1 The experimental setup for study of scattering in transmitted light [452]. The laser beam is broadened by the lens L_1, passes through the aperture diaphragm A_1 and is collected by the lens L_2. Sample S is illuminated by constant intensity light due to separating the central part from the broadening beam by the aperture diaphragm A_2. The diaphragms A_3 and A_4 isolate one "coherence area" (see also Section 5.2) for further detection using a PMT. The PMT exit signal falls on the correlator. In most cases the lens L_1 and the diaphragm A_1 can be omitted without any great decrease in quality.

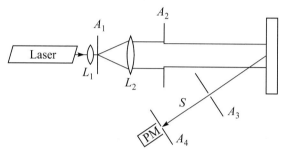

Figure A.2 The experimental setup for study of scattering in reflected light [452]. The allocation and designation of the scheme's elements is the same as for Fig. A.1.

cylindrical geometry could be combined using spherical optically dense samples [485, 486].

It was approximately at that time that the dynamic light scattering method using point sources and optical fiber detectors for backscattering arose and began to actively develop [570, 571]. Papers appeared, dedicated to one of the most complex areas of scattering multiplicity to interpret, intermediate scattering (see, e.g., [572–574] and references therein). There is no decrease in the intensity of research on non-Gaussian statistics of registered signals, especially for very low particle concentrations [415, 416, 423, 575]. A special, optical fiber, version of the homodyne photon correlation technique is applied in [415]. It should be added that, as the diffusion approximation supposes a vanishingly small light absorption by medium, it becomes unsuitable if this condition is not fulfilled. This problem can be partly solved by using the so-called telegraph equation instead of the diffusion equation [576, 577].

The description of the diffusing-wave spectroscopy fundamentals performed here is fairly brief for two reasons. Firstly, the description of multiple light scattering, independent of the experiment's geometry, is based on the principles of the diffusion approximation of the radiation-transport theory [347, 348, 354, 355] discussed in details previously (see Chapter 5). Secondly, vital details of the theory and of its various applications can be found in numerous articles published on this topic [441–463]. Let us now look at some specific applications of the photon correlation technique.

A.2
Method of Determination of the Mean Dimension and Concentration of Suspended Particles

It is not necessary to prove that the problem of determining particle dimensions is extremely relevant in practice. The existence of a comprehensive special literature dedicated to this problem shows this to be true [401–403, 578–581]. Also, all the literature cited in the previous paragraph, dedicated to diffusing-wave spectroscopy, is to a certain degree related to determining particle dimensions. Among the different methods used for this purpose in the micron- and nanosized range, the various optical methods cannot be beaten. But even among these, correlation spectroscopy is distinguished by its simplicity, precision, and convenience [319, 399–403, 582, 583]. It has now become a routine laboratory method for determining particle dimensions and the dynamics of the changes of the sizes in physics, chemistry, biology, industry, etc. Moreover, it is an absolute method which does not require any preliminary calibration [319, 399–403, 451, 538, 578–582].

However, all this only relates to infinitely diluted systems in which single scattering takes place and where Eq. (5.10) may be applied. To determine the particle dimensions directly in suspensions and highly turbid emulsions, such as milk, latexes, water emulsion paints, etc., this method is unsuitable due to the impossibility of guaranteeing single scattering. Taking this system with high extinction to low turbidity by dilution is not always possible or even justified, because of the physico-chemical features of dispersed systems, including particle dimensions, which can significantly change during such dilution. Moreover, the process of diluting such systems requires sampling, which makes it difficult to organize the operational and nondestructive control of their properties under working conditions.

Using the experimental and theoretical research on multiple light scattering in disperse systems with high extinction [141, 386, 390–394, 436–438], a method was developed [479, 480] which excluded these limitations and made it possible to include suspensions and highly turbid emulsions in the research using dynamic light scattering of materials. As a result, the efficiency of our mathematical model of multiple scattering (see Chapter 5) was confirmed and also the real possibility arose to carry out operational and nondestructive control of such materials *in situ* directly in working conditions. Thereby, it was possible to add correlation spectroscopy of multiple scattering to those instruments, which are not as simple as traditional ones but just as effective, used by engineers, technologists, and researchers. Let us have a look at the essence of this method which relates to scattering media optics and can be used to determine the mean particle dimension and concentration in suspensions and emulsions, applied in chemical technology, paint and varnish industry, food industry, etc.

The method suggested in [479, 480] was based, with this in mind, on scattering of monochromatic light on suspended particles, on separation of the scattered light beam and measurement of the half-width spectrum of scattered light in a separated beam. The regime used for light scattering in a dispersed system is the diffusion regime of multiple scattering. Moreover, the half-width spectrum is measured from, at least, two scattered light beams exiting the dispersed system through different parts of its border. The particle mean radius $\langle r \rangle$ and the concentration c are calculated using the already measured values of the half-width spectrum of scattered light. The essence of this method is based on the fact that, in the diffusion mode of multiple scattering, the half-width spectrum Γ_m (see Eqs. (5.10) and (5.61)) can be represented via Eq. (A.1)

$$\Gamma_m = \frac{k_B T}{6 \pi \overline{\eta} \langle r \rangle} \left(\frac{2 \pi n}{\lambda} \right)^2 \left[A + B \left(\frac{3 Q c}{4 \langle r \rangle} \right)^2 \right], \tag{A.1}$$

or, accounting for Eq. (5.26) via Eq. (A.2)

$$\Gamma_m = \frac{k_B T}{6 \pi \overline{\eta} \langle r \rangle} \left(\frac{2 \pi n}{\lambda} \right)^2 (A + B h^2), \tag{A.2}$$

where $\langle r \rangle$ is the mean radius of the suspended particles, dependent on the dimension and form of the scattering sample, and also on the relative position of the falling light beam, the sample and the optical system which separates light scattering beam (see Eq. (5.61)). The other symbols are the same as in Eqs. (5.10) and (5.26).

Furthermore, the measurement results of Γ_{m1} and Γ_{m2} in two separated beams will satisfy equations like Eq. (A.2)

$$\Gamma_{m1} = \frac{k_B T}{6 \pi \overline{\eta} \langle r \rangle} \left(\frac{2 \pi n}{\lambda} \right)^2 (A_1 + B_1 h^2), \tag{A.3}$$

$$\Gamma_{m2} = \frac{k_B T}{6 \pi \overline{\eta} \langle r \rangle} \left(\frac{2 \pi n}{\lambda} \right)^2 (A_2 + B_2 h^2). \tag{A.4}$$

The values of the parameters A_1, B_1 and A_2, B_2, which characterize the geometry of the used setup, can be calculated using the results of calibration measurements via Eqs. (A.5) and (A.6),

$$A_1 = \frac{\Gamma_{12} h_{01}^2 - \Gamma_{11} h_{02}^2}{\gamma_0 (h_{01}^2 - h_{02}^2)}, \quad B_1 = \frac{\Gamma_{11} - \Gamma_{12}}{\gamma_0 (h_{01}^2 - h_{02}^2)}, \tag{A.5}$$

$$A_2 = \frac{\Gamma_{22} h_{01}^2 - \Gamma_{21} h_{02}^2}{\gamma_0 (h_{01}^2 - h_{02}^2)}, \quad B_2 = \frac{\Gamma_{21} - \Gamma_{22}}{\gamma_0 (h_{01}^2 - h_{02}^2)}, \tag{A.6}$$

where

$$\gamma_0 \equiv \frac{k_B T_0}{6\pi r_0 \bar{\eta}_0} \left(\frac{2\pi n_0}{\lambda} \right)^2 . \tag{A.7}$$

Here Γ_{11}, Γ_{12}, Γ_{21}, and Γ_{22} are values of the half-width spectra of scattered light obtained by calibration measurements. For Γ_{ij} the first index corresponds to the separated light beam and the second to concentration. The zero index of all values in the formulas indicates that they have the characteristics of a standard dispersed system. Latter equations, Eqs. (A.5) and (A.6), are a consequence of Eqs. (A.3) and (A.4). The mentioned measurements should be carried out on a standard dispersed system with two well-known concentrations of suspended particles.

One of the peculiarities of the multiple scattering diffusion mode is the fact that the light beam exiting the dispersed system through the same part of its border has the same spectral half-width Γ_m for any angle value describing the direction of the diffused beam. Consequently, these equations, Eqs. (A.3) and (A.4), can be solved only if defined beams exit the dispersed system through different parts of its border. In this case the formulas for determining particle mean radius ($\langle r \rangle$) and volume concentration (c) will be

$$\langle r \rangle = \frac{k_B T}{6\pi \bar{\eta}} \left(\frac{2\pi n}{\lambda} \right)^2 \frac{B_1 A_2 - B_2 A_1}{B_1 \Gamma_2 - B_2 \Gamma_1}, \tag{A.8}$$

$$c = \frac{4 \langle r \rangle}{3Q} \sqrt{\frac{A_2 \Gamma_1 - A_2 \Gamma_2}{B_1 \Gamma_2 - B_2 \Gamma_1}}, \tag{A.9}$$

where Γ_1 and Γ_2 are the half-width spectra of multiple scattering in selected beams of scattered light.

An example of the implementation of this method is described in detail in [141, 480]. To determine the mean particle radius and concentration, polystyrene latex was used. The completely suspension-filled cuvette (diameter 20 mm, height 40 mm) was placed on the axis of the goniometer with an optical system for separating light beams, scattered by a dispersed system. The measurements were done in diffusion mode of multiple scattering. To do this it was necessary that the investigated dispersed system was quite turbid. It turned out that it was possible and necessary to control the level of turbidity by using the laser beam. Turbidity can be considered as sufficient to guarantee the diffusion mode of multiple scattering if the direct light beam does not exit the cuvette and fully scatters in it. All other methods of visually defining the necessary degree of turbidity are, as it seems, not effective enough.

The measurements were carried out in the following way. In two light beams, exiting the investigated system at angles of $90°$ and $120°$ referred to the incident beam the half-widths Γ_1 and Γ_2 were measured. Then, using Eq. (A.8) the mean radius of the particle $\langle r \rangle$ was calculated which allowed us, by using the tables [429, 430], to find the scattering efficiency factor of light scattered by suspended particles. Next, the volume concentration (c) was calculated using Eq. (A.9). The constants A_1, B_1 and A_2, B_2, as already mentioned, were determined by calibration measurements, fully identical to the basic ones, carried out using two standard dispersed systems, water suspensions with two different particle concentrations of the same monodisperse polystyrene latex. The particle size and dimensions in each system were known beforehand.

The values for particle radius and concentration in the investigated system, obtained by using Eqs. (A.8) and (A.9), were checked by independent measurements. Concentration was checked by evaporation and the radius by direct measurement with an electron microscope. The results agreed well, which confirms the efficiency of this method [480] and makes it possible to measure in the same apparatus particle dimension and concentration in highly turbid dispersed systems such as milk, latex water-emulsion paints, etc., without any prior dilution [479, 480]. In single scattering mode this is not at all possible as the spectral half-width Γ does not depend at all on the concentration of suspended particles.

A.3
Monitoring of Particle Motion in Drying Films

Using photon correlation technique it is possible to investigate the behavior of the particle diffusion coefficient, the viscosity of the dispersed medium and, on this basis, film forming kinetics. The simultaneous measuring of, apart from the correlation function, some other optical features of such colloidal systems creates additional possibilities. Below we shall demonstrate the results of one such complex investigation [584, 585].

In this research the chosen system was a concentrated water suspension of polystyrene particles of monodisperse latex with a radius of 100 nm. The sample's thickness (2 mm) was such that it could guarantee high scattering multiplicity. The sample, in the form of a drop, was placed on a object plate, fixed on the analytical balance beam and illuminated from below by a focussed laser beam. This construction made it possible to simultaneously measure the drop's mass and also its optical features, such as the multiple scattering spectrum half-width and the intensity of scattering light reflected back and going through the sample. The open top surface of the drop guaranteed free and continuous evaporation into the atmosphere. Changing of the drop's mass (see Fig. A.3, curve 4) made it possible to control the changes in the scatterer's concentration, determined as the mass of dry residue to mass of the drop ratio.

The experimental results are shown in Fig. A.3. It is clear that during the process of drying the drop on the background of monotonic concentration growth its optical features suffer sharp, like at a phase transition, changes. For the analysis it is convenient to split the experiments concentration range into three parts, 40–70%, 70–80%, and above 80%. In the first range, the sample's turbidity and all other optical features remain practically unchanged. On the border between the first and second parts, when the concentration is ~70%, there are dramatic changes in all controlled features. The decrease of Γ_m to zero shows that Brownian particle motion has stopped completely. This feature indicates the completion of

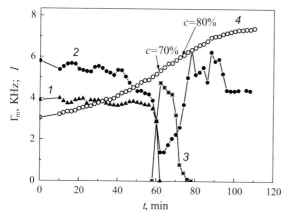

Figure A.3 Experimental results of the kinetic process of latex film drying. (1) is the half-width spectrum of multiple scattering, (2) intensity of backward scattered light, (3) intensity of transmitted light, (4) concentration of the scatterers.

an "order–disorder" structural phase transition, which results in the changes of the system into a quasiordered state with the densest packing. This is suggested by the value of the threshold concentration and the observed sample blooming on this part (curve 3), and also the results obtained using indirect methods in analogous systems [586]. When concentration is higher than 80% the system once more becomes turbid and oscillation intensity of backward scattering (curve 2) is determined by the film breaking due to the internal stresses.

So, this experiment shows once more that correlation spectroscopy of multiple scattering is a powerful instrument for the analysis of physico-chemical characteristics of highly turbid colloidal systems with the possibility of a quick control of technological processes.

A.4
Dynamics of Particle Formation and Growth

In this part we shall briefly look at the application of correlation spectroscopy in single scattering mode to investigate particle formation and growth in extremely unusual conditions: super-critical fluids, opaque systems, and sol–gel processes.

A.4.1
Supercritical Fluids

In recent years, supercritical fluids (SCF), which have high (like liquids) density, low (like gas) viscosity and intermediate (between liquid and gas) diffusion coefficient values, have become one of the most popular and promising media for implementing extraction and separation in polymer and chemical science and technology and also for creating new materials with specified properties (see, e.g., [506, 519, 538, 587, 589, 590]).

In [591–593], processes of particle formation and the dynamics of their growth have been observed in demixing processes of a solution of monodisperse polystyrene in *n*-butane at small variations of pressure which could change up to 30 MPa. Using light scattering the mean particle dimension was measured, as well.

A.4.2
Opaque Systems

Systems which have an absorption coefficient much higher than the scattering coefficient can be considered as opaque systems. The theoretical description of static and dynamic light scattering in such media (for example, oil) is not only a very important but also an extremely difficult task. The complex character of the refractive index and the unusual angular dependence of scattered light intensity in such systems creates further complications. The papers [594, 595] proposed a method to solve this problem, which consists in an analysis of light reflected from the substance's thin boundary layer (Fig. A.4). This makes it possible to avoid multiple scattering and obtain information on particle dimensions in opaque systems independent of their nature. In addition, the real and imaginary part of the medium's refractive index can be obtained in the same experiment.

This method, with the help of a correlator [408], was used to investigate the dispersivity of natural oils and the fractal structures in diffusion- and reaction-limited aggregation processes in both model and real systems (see, e.g., [594, 595]). The results are shown in Figs. A.5 and A.6. To the spectrum of research of opaque systems using traditional photon correlation techniques can also be added the rather successful attempts at studying the characteristics and processes taking place in human blood (see, e.g., [596, 597]).

A.4.3
Sol–Gel Process

Over the last few years interest in the sol–gel process has steadily grown. Understanding how the particles form and grow in this process is very important, not

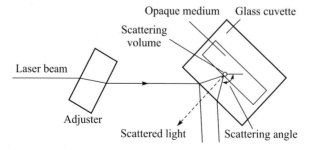

Figure A.4 The optical scheme of scattering in opaque media [594, 595].

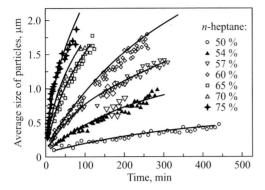

Figure A.5 Kinetics of asphaltene aggregation in a model system. Diffusion-limited mode [594, 595].

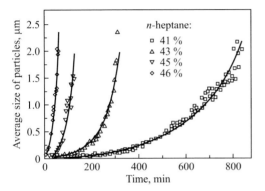

Figure A.6 Kinetics of asphaltene aggregation in a model system. Reaction-limited mode [594, 595].

just for practical chemical applications, but also for explaining the fundamentals of this process [598]. In the papers [599–601] dynamic light scattering within the limits of cylindrical geometry (see, e.g., Fig. 5.1) was applied for investigating the kinetics of the sol–gel process and dimension dynamics of the forming particles.

It was here that nanosized particle dimensions were first found of a few nanometers and growth dynamics was studied directly after hydrolysis–condensation of tetraisopropoxytitan (Fig. A.7). The obtained dependence of $R(t)$ and $I(t)$ (see Fig. A.7) made it possible to determine the value of the fractal dimension of particles as $d_f = 1.4 \pm 0.1$ and to draw a conclusion concerning their rather loose structure (for dense particles, according to the fractal dimension definition, this is $d_f = 3$). This experiment [599–601], apart from everything else, is remarkable that it demonstrates the possibility of an effective applicability of the correlation spectroscopy for needs of a rapidly developing technology with application to nanostructures.

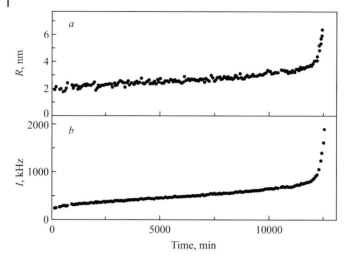

Figure A.7 Dependence of the mean radius of the particle R (a) and the intensity of scattered light I (b) on time
[599, 600].

In conclusion, we can say that correlation spectroscopy has become a powerful
and convenient tool for the analysis of disperse systems, including very turbid ones,
for express control of technological processes *in situ* and also critical phenomena.
The extension of this method's possibilities will, it seems, continue for some time
yet. As for the multiple scattering spectra theory, the diffusion approximation
of the radiation-transport theory describes the main features and essence of
the phenomenon very well. Nevertheless, there are still many questions, related
to intermediate scattering multiplicity, scattering in strongly interacting particle
systems, etc., which are still waiting for their solution.

References

1 V. L. Ginzburg, In: *Physics in the XXth Century: Development and Progress* (Nauka, Moscow, **1984**, pp. 281–330).

2 D. Yu. Ivanov, *Dokl. Phys.* **47**, 267 (**2002**) [*Dokl. Akad. Nauk* **383**, 478 (2002)].

3 R. Feynman, *The Character of Physical Law* (Cox and Wyman, London, **1965**; Mir, Moscow, **1968**).

4 A. Einstein, *Collected Scientific Works* (Nauka, Moscow, **1969**, vol. 3, p. 432).

5 *Critical Phenomena.* In: *Proceedings of Conference on Phenomena in the Neighborhood of Critical Points* (Edited by M. S. Green and J. V. Sengers, NBS Misc. Publ. 273, Washington **1966**, pp. 1–242).

6 G. E. Uhlenbeck, *The classical theories of the critical phenomena.* In: *Proceedings of the Conference on Phenomena in the Neighborhood of Critical Points* (Edited by M. S. Green and J. V. Sengers, NBS Misc. Publ. 273, Washington, **1966**, pp. 3–6).

7 T. Andrews, *Phil. Trans. Royal Soc. London* **159**, 575 (**1869**).

8 T. Andrews, *Phil. Trans. Royal Soc. London* **166**, 421 (**1876**).

9 T. Andrews, *On the Continuity of the Gaseous and Liquid States of Matter* (Gostekhteorizdat, Moscow, Leningrad, **1933**).

10 D. I. Mendeleev, *Chemical Journal* (Edited by N. Sokolov and A. Engelhardt **3**, 81 (**1860**).

11 D. I. Mendeleev, *Poggendorfs Annalen der Physik* **141**, 618 (**1870**).

12 C. Cagniard de La Tour, *Ann. Chim. et Phys.* **21**, 127, 178 (**1822**).

13 M. Gliozzi *Storia della Fisica* (Storia delle scienze, Torino, **1965**; Mir, Moscow, **1970**).

14 J. D. van der Waals, Ph.D. thesis (Sijthoff, Leiden, **1873**).

15 J. S. Rowlinson, *J. D. van der Waals: On the continuity of the gaseous and liquid states.* In: *Studies in Statistical Mechanics* **XIV** (Edited by J. L. Lebowitz, North-Holland, Amsterdam, **1988**).

16 J. Hopkinson, *Philos. Trans. R. Soc,* 443 (**1889**).

17 P. Weiss, *J. Phys. Rad.* **6**, 667 (**1907**).

18 G. Tammann, *Zeitschrift Anorg. Chem.* **107**, 1 (**1919**).

19 C. H. Johansson, J. O. Linde, *Ann. Physik* **78**, 439 (**1925**).

20 W. L. Bragg and E. J. Williams, *Proc. R. Soc. London A* **145**, 699 (**1934**).

21 W. L. Bragg and E. J. Williams, *Proc. R. Soc. London A* **151**, 540 (**1935**).

22 E. J. Williams, *Proc. R. Soc. London A* **152**, 231 (**1935**).

23 W. Gorsky, *Zeitschrift für Physik* **50**, 64 (**1928**).

24 M. Kac, G. E. Uhlenbeck, and P. C. Hemmer, *J. Math. Phys.* **4**, 216, 229 (**1963**); *ibid* **5**, 60 (**1964**).

25 L. D. Landau, *Collected Works* (Nauka, Moscow, **1969**, vols. 1 and 2).

26 H. E. Stanley, *Introduction to Phase Transitions and Critical Phenomena* (Clarendon Press, Oxford, **1971**; Mir, Moscow, **1973**).

27 L. D. Landau and E. M. Lifshitz, *Statistical Physics* (Nauka, Moscow,

Critical Behavior of Nonideal Systems. Dmitry Yu. Ivanov
Copyright © 2008 WILEY-VCH Verlag GmbH & Co. KGaA, Weinheim
ISBN: 978-3-527-40658-6

1976; Academy of Sciences Publishing House, Berlin, **1987**).

28 R. Gouy, *Compt. Rend. Acad. Sci. Paris* **115**, 720 (**1892**).

29 R. Gouy, *Compt. Rend. Acad. Sci. Paris* **116**, 1289 (**1893**).

30 A. G. Stoletov, *Collected Works* in 4 volumes (Gostekhteorizdat, Moscow, Leningrad, **1939**).

31 A. G. Stoletov, *Collected Works* in 4 volumes (Gostekhteorizdat, Moscow, Leningrad, **1939**) **1**, 276, 307, 333.

32 M. A. Weinberger and W. G. Schneider, *Can. J. Chem.* **30**, 422 (**1952**).

33 M. A. Weinberger, H. W. Habgood, and W. G. Schneider, *Can. J. Chem.* **30**, 815 (**1952**).

34 M. A. Weinberger and W. G. Schneider, *Can. J. Chem.* **30**, 847 (**1952**).

35 M. von Smoluchowski, *Ann. d. Physik. (Leipzig)* **25**, 205 (**1908**).

36 L. S. Ornstein and F. Zernike, *Proc. Sec. Sci. Kon. Akad. Wetensch. Amsterdam* **17**, 793 (**1914**).

37 L. Onsager, *Phys. Rev.* **65**, 117 (**1944**).

38 S. G. Brush, *Rev. Mod. Phys.* **39**, 883 (**1967**).

39 M. A. Anisimov, *Uspekhi Fiz. Nauk* **114**, 249 (**1974**).

40 M. A. Anisimov, *Critical Phenomena in Liquids and Liquid Crystals* (Nauka, Moscow, **1987**; Gordon and Breach, New York, **1992**).

41 M. E. Fisher, *The Nature of Critical Points* (University of Colorado Press, Boulder, Colorado, USA, **1965**; Mir, Moscow, **1968**).

42 E. A. Guggenheim, *J. Chem. Phys.* **13**, 253 (**1945**).

43 A. Michels, B. Blaisse, and C. Michels, *Proc. Roy. Soc. A* **160**, 358 (**1937**).

44 J. S. Kouvel and M. E. Fisher, *Phys. Rev. A* **136**, 1626 (**1964**).

45 J. E. Verschaffelt, *Proc. Sec. Sci. Kon. Akad. Wetenschappen* **2**, 588 (**1900**).

46 J. M. H. Levelt-Sengers, *Physica A* **82**, 319 (**1976**).

47 *Phase Transitions and Critical Phenomena* (Edited by C. Domb and M. S. Green, Academic Press, London, New York, **1972**, vol. 1, pp. vii–viii).

48 L. P. Kadanoff, W. Götze, D. Hamblen, R. Hecht, E. A. S. Levis, V. V. Palciauskas, M. Rayl, J. Swift, D. Aspnes, and J. Kane, *Rev. Mod. Phys.* **39**, 395 (**1967**).

49 J. M. H. Levelt Sengers, W. L. Greer, and J. V. Sengers, *J. Phys. Chem. Ref. Data* **5** (1), 1 (**1976**).

50 J. Zinn-Justin, *Quantum Field Theory and Critical Phenomena* (Clarendon Press, Oxford, **1993**).

51 A. N. Vasil'iev, *The Field Theoretic Renormalization Group in Critical Behavior Theory and Stochastic Dynamics* (Chapman and Hall/CRC, Boca Raton, **2004**; PINF Publ., St. Petersburg, **1998**).

52 Shang-keng Ma, *Modern Theory of Critical Phenomena* (W. A. Benjamin, London, **1976**; Mir, Moscow, **1980**).

53 R. Brout, *Phase Transitions* (W. A. Benjamin, New York, **1965**).

54 M. E. Fisher, *The theory of critical point singularities*. In: *Critical Phenomena, International School of Physics "Enrico Fermi"*, Lecture Course (Edited by M. S. Green, Academic Press, New York, **1971**, pp. 1–124).

55 A. Z. Patashinsky and V. L. Pokrovsky, *Fluctuation Theory of Phase Transitions* (Pergamon, Oxford, **1979**; Nauka, Moscow, **1975**).

56 *Physics of Simple Liquids: Statistical Theory* (Edited by H. N. V. Temperley, J. S. Rowlinson, and G. S. Rushbrooke, North-Holland, Amsterdam, **1968**; Mir, Moscow, **1971**).

57 M. E. Fisher, *J. Math. Phys.* **5**, 944 (**1964**).

58 B. Widom, *J. Chem. Phys.* **43**, 3898 (**1965**).

59 C. Domb and D. L. Hunter, *Proc. Phys. Soc. London* **86**, 1147 (**1965**).

60 A. Z. Patashinsky and V. L. Pokrovsky, *Zhurn. Eksper. Teor. Fiz.* **50**, 439 (**1966**).

61 L. P. Kadanoff, *Physics* **2**, 263 (**1966**).

62 L. P. Kadanoff, *Scaling, universality and operator algebras*. In: *Phase Transitions and Critical Phenomena* (Edited by C. Domb and M. S. Green. Academic Press, London, New York, **1976**, vol. 5a, pp. 2–34).

63 F. J. Wegner, *Phys. Rev. B* **5**, 4529 (**1972**).

64 F. J. Wegner, *The critical state, general aspects.* In: *Phase Transitions and Critical Phenomena* (Edited by C. Domb and M. S. Green. Academic Press, London, New York, **1976**, vol. 6, pp. 7–124).

65 *Physics of Simple Liquids: Experiment* (Edited by H. N. V. Temperley, J. S. Rowlinson, and G. S. Rushbrooke, North-Holland, Amsterdam, **1968**; Mir, Moscow, 1971).

66 P. Heller, *Rep. Progr. Phys.* **30**, 731 (**1967**).

67 A. Aharony, *Dependence of universal critical behaviour on symmetry and range of interaction.* In: *Phase Transitions and Critical Phenomena* (Edited by C. Domb and M. S. Green. Academic Press, London, New York, **1976**, vol. 6, pp. 357–424).

68 V. K. Semenchenko, *Zhurn. Fiz. Khimii* **21**, 1461 (**1947**).

69 V. P. Skripov, *Voprosy Istorii Estestvoznaniya i Tehniki* **4**, 119 (**1995**).

70 V. L. Ginzburg, *Fiz. Tverd. Tela (Leningrad)* **2**, 2031 (**1960**) [Sov. Phys. Solid State **2**, 1824 (1960)].

71 K. Wilson and J. Kogut, *The Renormalization Group and the ε-Expansion* (Wiley, New York, **1974**; Mir, Moscow, **1975**); *Phys. Rep.* **12C** (2), 75 (**1974**).

72 K. Wilson, *Phys. Rev. B* **4**, 3174, 3184 (**1971**).

73 K. G. Wilson and M. E. Fisher, *Phys. Rev. Lett.* **28**, 240 (**1972**).

74 E. Brézin, J. C. LeGuillou, and J. Zinn-Justin, *Field theoretical approach to critical phenomena.* In: *Phase Transitions and Critical Phenomena* (Edited by C. Domb and M. S. Green, Academic Press, London, New York, **1976**, vol. 6, pp. 125–246).

75 E. Brézin, *Critical behaviour from the field theoretical renormalization group techniques.* In: *Phase Transitions: Cargèse 1980* (Edited by M. Lévy, J. C. Le Guillou, and J. Zinn-Justin, Plenum Press, New York, London, **1982**, pp. 331–348).

76 A. Aharony and P. C. Hohenberg, *Phys. Rev. B* **13**, 3081 (**1976**).

77 G. A. Baker Jr. and D. I. Meiron, *Phys. Rev. B* **17**, 1365 (**1978**).

78 A. Onuki and K. Kawasaki, *Ann. Phys.* **121**, 456 (**1979**).

79 V. P. Skripov and V. K. Semenchenko, *Zhurn. Fiz. Khimii* **29**, 174 (**1955**).

80 M. N. Bagacky, A. V. Voronel, and V. G. Gusak, *Pis'ma v Zhurn. Eksper. Teor. Fiz.* **43**, 728 (**1962**).

81 H. I. Amirhanov, I. G. Gurvich, and E. V. Matizen, *Dokl. Akad. Nauk SSSR* **100**, 735 (**1955**).

82 V. P. Skripov and Yu. D. Kolpakov, *Optika i Spektroskopiya* **19**, 392 (**1965**); *ibid*, 616.

83 L. M. Artyuhovskaya, E. T. Shimanskaya, and Yu. I. Shimansky, *Zhurn. Eksper. Teor. Fiz.* **63**, 2159 (**1972**).

84 L. A. Makarevich and O. N. Sokolova, *Zhurn. Fiz. Khimii* **57**, 763 (**1973**).

85 I. R. Krichevsky, *Phase Equilibria in Solutions at High Pressures* (Goskhimizdat, Moscow, Leningrad, **1952**).

86 J. M. H. Levelt Sengers, A. Harvey, R. Crovetto, and J. S. Gallagher, *Fluid Phase Equilib.* **81**, 85 (**1992**).

87 D. Yu. Ivanov, L. A. Makarevich, and O. N. Sokolova, *JETP Lett.* **20**, 121 (**1974**) [Pis'ma v Zh. Eksp. Teor. Fiz. **20**, 272 (1974)].

88 D. Yu. Ivanov and L. A. Makarevich, *Dokl. Akad. Nauk SSSR* **220**, 1103 (**1975**).

89 D. Yu. Ivanov and V. K. Fedyanin, Preprint No. P4-8430 JINR (Joint Institute for Nuclear Research, Dubna, Russia, **1974**).

90 D. Yu. Ivanov and V. K. Fedyanin, Preprint No. P4-8429 JINR (Joint Institute for Nuclear Research, Dubna, Russia, **1974**).

91 D. Yu. Ivanov, Ph.D. Dissertation (Leningrad State University, Leningrad, **1975**).

92 D. Yu. Ivanov and L. A. Makarevich, *Behaviour of pure matter compressibility near a critical point* ($\Delta\rho = 0$, $\tau > 0$). In: *Liquid State Physics*

(Kiev University Publishing House, Kiev, **1977**, N 5, pp. 28–31).

93 D. Beysens, *Status of the experimental situation in critical binary fluids*. In: *Phase Transitions: Cargèse 1980* (Edited by M. Lévy, J. C. Le Guillou, and J. Zinn-Justin, Plenum, New York, **1982**, pp. 25–62).

94 D. Beysens, A. Bourgou, and P. Calmettes, *Phys. Rev. A* **26**, 3589 (**1982**).

95 D. Beysens, M. Gbadamassi, and L. Boyer, *Phys. Rev. Lett.* **43**, 1253 (**1979**).

96 D. Beysens, *New critical behavior induced by shear flow in binary fluids*. In: *Scattering Techniques Applied to Supramolecular and Non-equilibrium Systems* (Edited by S. H. Chen, B. Chu, and R. Nossal, Plenum, New York, **1981**).

97 D. Beysens, M. Gbadamassi, and B. Monsef-Bouanz, *Phys. Rev. A* **28**, 2491 (**1983**).

98 W. Wagner, N. Kurzeja, and B. Pieperbeck, *Fluid Phase Equilb.* **79**, 151 (**1992**).

99 N. Kurzeja, Th. Tielkes, and W. Wagner, *Int. J. Thermophys.* **20**, 531 (**1999**).

100 R. P. Feynman, *QED: The Strange Theory of Light and Matter* (Princeton University Press, Princeton, NJ, **1985**; Nauka, Moscow, **1988**).

101 M. E. Fisher: *Notes, definitions, and formulas for critical point singularities*. In: *Proceedings of Conference on Phenomena in the Neighborhood of Critical Points* (Edited by M. S. Green and J. V. Sengers, NBS Misc. Publ. 273, Washington **1966**, pp. 21–25).

102 J. C. Le Guillou and J. Zinn-Justin, *Phys. Rev. Lett.* **39**, 95 (**1977**).

103 J. C. Le Guillou and J. Zinn-Justin, *Phys. Rev. B* **21**, 3976 (**1980**).

104 G. Ahlers, *Static and dynamic critical phenomena near the superfluid transition in* ^{4}He. In: *Phase Transitions: Cargèse 1980* (Edited by M. Lévy, J. C. Le Guillou, and J. Zinn-Justin, Plenum Press, New York, London, **1982**, pp. 1–24).

105 G. A. Baker, Jr., *What is series extrapolation about?* In: *Phase Transitions: Cargèse 1980* (Edited by M. Lévy, J. C. Le Guillou, and J. Zinn-Justin, Plenum Press, New York, London, **1982**, pp. 149–152).

106 B. G. Nickel, *The problem of confluent singularities*. In: *Phase Transitions: Cargèse 1980* (Edited by M. Lévy, J. C. Le Guillou, and J. Zinn-Justin, Plenum Press, New York, London, **1982**, pp. 291–324).

107 J. J. Rehr, *Differential approximants and confluent singularity analysis*. In: *Phase transitions: Cargèse 1980* (Edited by M. Lévy, J. C. Le Guillou, and J. Zinn-Justin, Plenum Press, New York, London, **1982**, pp. 325–330).

108 A. Aharony and G. Ahlers, *Phys. Rev. Lett.* **44**, 782 (**1980**).

109 *Physical Encyclopaedia*, (Grand Russian Encyclopaedia, Moscow, **1992**, vol. 3, p. 672).

110 L. A. Makarevich, Ph.D. Dissertation (Moscow, **1967**).

111 L. A. Weber and D. R. Defibaugh, *Fluid Phase Equilib.* **150–151**, 731 (**1998**).

112 L. A. Weber, *Phys. Rev. A* **2**, 2379 (**1970**).

113 R. H. Wentorf, Jr., *J. Chem. Phys.* **24**, 607 (**1956**).

114 A. Oda, M. Uematsu, and K. Watanabe, *Bull. JSME* **26**, 1590 (**1983**).

115 P. C. Hohenberg and M. Barmatz, *Phys. Rev. A* **6**, 289 (**1972**).

116 L. A. Makarevich, E. S. Sokolova, and G. A. Sorina, *Zhurn. Fiz. Khimii* **42**, 22 (**1968**).

117 B. Chu, *Experiments on the critical opalescence of binary liquid mixtures: Elastic scattering*. In: *Proceedings of the Conference on Phenomena in the Neighborhood of Critical Points* (Edited by M. S. Green and J. V. Sengers, NBS Misc. Publ. 273, Washington, **1966**, pp. 123–129).

118 L. Cailletet and E. Mathias, *Compt. Rend.* **102**, 1202 (**1886**); *ibid.* **104**, 1563 (**1887**).

119 B. Widom and J. S. Rowlinson, *J. Chem. Phys.* **52**, 1670 (**1970**).

120 B. Widom and F. H. Stillinger, *J. Chem. Phys.* **58**, 616 (**1973**).

121 A. T. Berestov, E. E. Gorodecky, and V. M. Zaprudsky, *Pis'ma v Zhurn. Eksper. Teor. Fiz.* **21**, 56 (**1975**).

122 R. E. Goldstein and N. W. Ashcroft, *Phys. Rev. Lett.* **55**, 2164 (**1985**).

123 R. E. Goldstein, A. Parola, N. W. Ashcroft, M. W. Pestak, M. H. W. Chan, J. R. de Bruyn, and D. A. Balzarini, *Phys. Rev. Lett.* **58**, 41 (**1987**).

124 R. E. Goldstein and A. Parola, *Acc. Chem. Res.* **22**, 77 (**1989**).

125 A. Kumar, H. R. Krishnamurthy, and E. S. R. Gopal, *Phys. Rep.* **98**, 57 (**1983**).

126 D. Yu. Ivanov, *Gravity effects on pure matter properties near the critical point.* In: *Thermophysical Properties of Substances: Proceedings of VIII All-Union Conference* (Novosibirsk, Moscow, **1989**, vol. 1, pp. 36–43).

127 V. K. Fedjanin and D. Yu. Ivanov, *Gravity effects on the properties of pure matter near the critical point.* In: *Abstracts of Xth IUPAC Conference on Chemical Thermodynamics* (Prague, Czechoslovakia, **1988**, A 27).

128 D. Yu. Ivanov, *(P-V-T)-study near the critical point of a pure substance.* In: *High Pressure Science and Technology* (Edited by B. Vodar and Ph. Marteau, Pergamon, Oxford, UK, **1980**, pp. 713–714).

129 S. Jüngst, B. Knuth, and F. Hensel, *Phys. Rev. Lett.* **55**, 2160 (**1985**).

130 S. C. Greer, *Phys. Rev. A* **14**, 1770 (**1976**).

131 S. C. Greer, B. K. Das, A. Kumar, and E. S. R. Gopal, *J. Chem. Phys.* **79**, 4545 (**1983**).

132 H. A. Kierstead, *Phys. Rev. A* **3**, 329 (**1971**).

133 H. A. Kierstead, *Phys. Rev. A* **7**, 242 (**1973**).

134 J. E. Mayer and M. G. Mayer, *Statistical Mechanics* (John Wiley and sons, Inc., New York, **1940**).

135 D. S. Cannel, *Phys. Rev. A* **12**, 225 (**1975**).

136 M. W. Pestak and M. H. W. Chan, *Phys. Rev. B* **30**, 274 (**1984**).

137 E. Carlon and A. Drzewiski, *Fluid Phase Equilib.* **150**, 583 (**1998**).

138 D. Beysens and P. Calmettes, *J. Chem. Phys.* **66**, 766 (**1977**).

139 D. Beysens and J. Wesfreid, *J. Chem. Phys.* **71**, 119 (**1979**).

140 L. D. Landau and E. M. Lifshitz, *Continuum Electrodynamics* (Nauka, Moscow, **1982**).

141 D. Yu. Ivanov, Doctoral Dissertation in Mathematical Physics (St. Petersburg State University, **2001**).

142 D. Yu. Ivanov, *Macroscopic field influence on the critical exponents.* In: *Proceedings of the 14th European Conference on Thermophysical Properties* (Lyon, Villeurbanne, France, **1996**, p. 463).

143 I. K. Kamilov and Kh. K. Aliev, *Statistical Critical Phenomena in Magneto-Ordered Crystals* (Izdatelstvo Dagestan Nauchn. Center Ross. Akademii Nauk, Makhachkala, **1993**).

144 J. Weiner, K. H. Langley, and N. C. Ford, Jr., *Phys. Rev. Lett.* **32**, 879 (**1974**).

145 M. R. Moldover and R. W. Gammon, *J. Chem. Phys.* **80**, 528 (**1984**).

146 A. A. Bulavin and Yu. I. Shimansky, *Pis'ma v Zhurn. Eksper. Teor. Fiz.* **29**, 482 (**1979**).

147 A. A. Bulavin, P. G. Ivanicky, A. N. Maistrenko, Yu. B. Mel'nichenko, and Yu. I. Shimansky, *Ukr. Fiz. Zhurn.* **23**, 1125 (**1978**).

148 M. Ley-Koo and M. S. Green, *Phys. Rev. A* **16**, 2483 (**1977**).

149 J. F. Nicoll and P. C. Albricht, *Renormalization-group approach to critical phenomena in fluids: A theory of the coexistence curve diameter.* In: *Proceedings of the 8th Symposium on Thermophysical Properties* (Gaithersburg, Maryland, **1981**, pp. 377–382).

150 J. F. Nicoll, *Phys. Rev. A* **24**, 2203 (**1981**).

151 B. J. Thijsse, *J. Chem. Phys.* **74**, 4678 (**1981**).

152 J. M. H. Levelt Sengers and W. T. Chen, *J. Chem. Phys.* **56**, 595 (**1972**).

153 J. Garland and C. W. Thoen, *Phys. Rev. A* **13**, 1601 (**1976**).

154 J. M. H. Levelt Sengers, J. Straub, and M. Vicentini-Missoni, *J. Chem. Phys.* **54**, 5034 (**1971**).

155 G. W. Mulholland and J. J. Rehr, *J. Chem. Phys.* **60**, 1297 (**1974**).

156 J. J. Rehr and N. D. Mermin, *Phys. Rev. A* **8**, 472 (**1973**).

157 J. Hubbard and P. Schofield, *Phys. Lett. A* **40**, 245 (**1972**).

158 M. Ley-Koo and M. S. Green, *Phys. Rev. A* **23**, 2650 (**1981**).

159 H. W. Habgood and W. G. Schneider, *Can. J. Chem.* **32**, 164 (**1954**).

160 H. W. Habgood and W. G. Schneider, *Can. J. Chem.* **32**, 98 (**1954**).

161 B. Wallace and H. Meyer, *Phys. Rev. A* **2**, 1563 (**1970**).

162 B. Wallace and H. Meyer, *Phys. Rev. A* **2**, 1610 (**1970**).

163 L. A. Makarevich, O. N. Sokolova, and A. M. Rozen, *Zh. Eksp. Teor. Fiz.* **67**, 615 (**1974**) [Sov. Phys. JETP **40**, 305 (**1974**)].

164 D. Yu. Ivanov, *Dokl. Phys.* **49**, 73 (**2004**) [Dokl. Akad. Nauk **394**, 757 (2004)]

165 D. Yu. Ivanov, *Model of the critical behavior of real systems*. In: *Non-linear Dielectric Phenomena in Complex Liquids* (Edited by S. J. Rzoska and V. P. Zhelezny, Kluwer, The Netherlands, **2004**, pp. 153–161) [*NATO Science Series. Series II: Mathematics, Physics and Chemistry –vol. 157*].

166 R. Balescu, *Equilibrium and Non-equilibrium Statistical Mechanics*, vol. 1 (Wiley, New York, London, Sydney, Toronto, **1975**; Mir, Moscow, **1978**).

167 Ya. I. Frenkel, *Zhurn. Eksper. Teor. Fiz.* **9**, 952 (**1939**); *Kinetic Theory of Liquids* (Akademia Nauk, Moscow, Leningrad, **1945**).

168 R. Evans and U. M. B. Marini, *J. Chem. Phys.* **86**, 7138 (**1987**).

169 R. Evans, *J. Phys.: Condens. Matter.* **2**, 8989 (**1990**).

170 M. Thommes and G. H. Findenegg, *Langmuir* **10**, 4270 (**1994**).

171 R. B. Griffiths, *Rigorous results and theorems*. In: *Phase Transitions and Critical Phenomena* (Edited by C. Domb and M. S. Green, Academic Press, London, New York, **1976**, vol. 1, pp. 7–109).

172 A. A. Licalter, *Uspekhi Fiz. Nauk* **170**, 831 (**2000**).

173 D. Wirtz, D. E. Werner, and G. G. Fuller, *J. Chem. Phys.* **101**, 1679 (**1994**).

174 P. Guenoun, F. Perrot, and D. Beysens, *Phys. Rev. A* **38**, 2588 (**1988**).

175 L. A. Bulavin, A. V. Oleinikova, and A. V. Petrovitskij, *Int. J. Thermophys.* **17**, 137 (**1996**).

176 R. R. Singh and K. S. Pitzer, *J. Chem. Phys.* **92**, 6775 (**1990**).

177 T. Narayanan and K. S. Pitzer, *Int. J. Thermophys.* **15**, 1037 (**1994**).

178 S. Wiegand, M. Kleemeier, J.-M. Schröder, W. Schröer, and H. Weingärtner, *Int. J. Thermophys.* **15**, 1045 (**1994**).

179 P. Schofield, E. J. Litster, and T. J. Ho, *Phys. Rev. Lett.* **23**, 1098 (**1969**).

180 A. V. Chaly and L. M. Chernenko, *Zhurn. Eksper. Teor. Fiz.* **87**, 187 (**1984**).

181 J. V. Sengers and J. M. J. van Leeuwen, *Int. J. Thermophys.* **6**, 545 (**1985**).

182 J. H. Sikkenk, J. M. J. van Leeuwen, and J. V. Sengers, *Physica A* **139**, 1 (**1986**).

183 J. V. Sengers and J. M. H. Levelt Sengers, *Ann. Rev. Phys. Chem.* **37**, 189 (**1986**).

184 J. S. Rowlinson and B. Widom, *Molecular Theory of Capillarity* (Clarendon, Oxford, **1982**).

185 M. R. Moldover, J. V. Sengers, R. W. Gammon, and R. J. Hocken, *Rev. Mod. Phys.* **51**, 79 (**1979**).

186 A. V. Voronel, *Zhurn. Fiz. Khimii* **35**, 958 (**1961**).

187 Yu. P. Blagoi and V. G. Gusak, *Zhurn. Eksper. Teor. Fiz.* **56**, 592 (**1969**).

188 J. Straub, R. Lange, K. Nitsche, and K. Kemmerle, *Int. J. Thermophys.* **7**, 343 (**1986**).

189 J. Straub and K. Nitsche, *Fluid Phase Equilib.* **88**, 183 (**1993**).

190 J. Straub, A. Haupt, and L. Eicher, *Int. J. Thermophys.* **16**, 1033 (**1995**).

191 H. M. Barmatz, I. Hahn, J. A. Lipa, and R. V. Duncan, *Rev. Mod. Phys.*, **79**, 1 (**2007**).

192 S. V. Krivokhizha, O. A. Lugovaya, I. L. Fabelinskii, et al., *Zh. Eksp. Teor. Fiz.* **103**, 115 (**1993**) [JETP **76**, 62 (**1993**)].

193 A. A. Sobyanin, *Uspekhi Fiz. Nauk* **149** (2), 325 (**1986**).

194 A. A. Povodyrev, G. X. Jin, S. B. Kiselev, and J. V. Sengers, *Int. J. Thermophys.* **17**, 909 (**1996**).

195 V. G. Puglielli and N. C. Ford (Jr.), *Phys. Rev. Lett.* **25**, 143 (**1970**).

196 M. Corti and V. Degiorgio, *Phys. Rev. Lett.* **45**, 1045 (**1980**).

197 M. Corti, V. Degiorgio, and M. Zulauf, *Phys. Rev. Lett.* **48**, 1617 (**1982**).

198 G. Dietler and D. Cannell, *Phys. Rev. Lett.* **60**, 1852 (**1988**).

199 K. Hamano, N. Kuwahara, T. Koyama, and S. Harada, *Phys. Rev. A* **32**, 3168 (**1985**).

200 M. E. Fisher, *Phys. Rev. Lett.* **57**, 1911 (**1986**).

201 C. Bagnuls and C. Bervillier, *Phys. Rev. Lett.* **58** (5), 435 (**1987**).

202 L. Reatto, *Nuovo Cimento D* **8**, 497 (**1986**).

203 A. I. Rusanov, *Micelle Formation in Surfactant Solutions* (Khimia, St. Petersburg, **1992**).

204 K. Hamano et al., *Int. J. Thermophys.* **10**, 389 (**1989**).

205 M. E. Fisher and H. Nakanishi, *J. Chem. Phys.* **75**, 5857 (**1981**).

206 *Finite-Size Scaling and Numerical Simulation of Statistical Systems* (Edited by V. Privman, World Scientific, Singapore, **1990**).

207 M. Krech and S. Dietrich, *Phys. Rev. Lett.* **66**, 345 (**1991**).

208 K. K. Mon and K. Binder, *Phys. Rev. E* **48**, 2498 (**1993**).

209 P. A. Rikvold, B. M. Gorman, and M. A. Novotny, *Phys. Rev. E* **47**, 1474 (**1993**).

210 A. Z. Panagiotopoulos, *Int. J. Thermophys.* **15**, 1057 (**1994**).

211 H.-P. Deutsch and K. Binder, *J. de Physique* **3**, 1049 (**1993**).

212 A. R. Muratov, *Zhurn. Eksper. Teor. Fiz.* **120** (1), 104 (**2001**).

213 J. Jacob et al., *Phys. Rev. E* **58**, 2200 (**1998**).

214 V. A. Agayan, M. A. Anisimov, and J. V. Sengers, *Phys. Rev. E* **64**, 026125 (**2001**)

215 M. A. Anisimov, *J. Phys. A. Condens. Matter* **12**, 451 (**2000**).

216 Y. C. Kim, M. A. Anisimov, J. V. Sengers, and E. Luijten, *J. Stat. Phys.* **110**, 591 (**2003**).

217 D. Yu. Ivanov, *Principle of corresponding states and Ginzburg criterion.* In: *Abstracts of the XVIth International Conference on Chemical Thermodynamics* (Suzdal, Moscow, **2007**, vol. 1, pp. 91–92).

218 P. Novak, R. Kleinrahm, and W. Wagner, *J. Chem. Thermodyn.* **28**, 1441 (**1996**).

219 R. Gilgen, R. Kleinrahm, and W. Wagner, *J. Chem. Thermodyn.* **26**, 399 (**1994**).

220 R. Kleinrahm and W. Wagner, *J. Chem. Thermodyn.* **18**, 739 (**1986**).

221 R. Gilgen, R. Kleinrahm, and W. Wagner, *J. Chem. Thermodyn.* **24**, 953 (**1992**).

222 P. Novak, R. Kleinrahm, and W. Wagner, *J. Chem. Thermodyn.* **28**, 1423 (**1996**).

223 R. Gilgen, R. Kleinrahm, and W. Wagner, *J. Chem. Thermodyn.* **26**, 383 (**1994**).

224 R. Gilgen, R. Kleinrahm, and W. Wagner, *J. Chem. Thermodyn.* **24**, 1243 (**1992**).

225 D. Yu. Ivanov, *On the nature of the critical point.* In: *Proceedings of the International Conference: Phase Transition, Critical and Non-Linear Phenomena in Condensed Media.* (Makhachkala, Russia, **2007**, pp. 284–287).

226 D. Yu. Ivanov, *Dokl. Phys.* **52**, 380 (**2007**) [Dokl. Akad. Nauk **415**, 330 (**2007**)].

227 M. Funke, R. Kleinrahm, and W. Wagner, *J. Chem. Thermodyn.* **34**, 2001 (**2002**).

228 P. Claus, R. Kleinrahm, and W. Wagner, *J. Chem. Thermodyn.* **35**, 159 (**2003**).

229 M. S. Green, *Introduction*. In: *Proceedings of the Conference on Phenomena in the Neighborhood of Critical Points* (Edited by M. S. Green and J. V. Sengers, NBS Misc. Publ. 273, Washington **1966**, pp. ix–xi).

230 E. S. Wu and W. W. Webb, *Phys. Rev. A* **8**, 2065 (**1973**); *ibid*, 2077.

231 M. C. Chang and A. Houghton, *Phys. Rev. Lett.* **44**, 785 (**1980**).

232 C. Bervillier, *Phys. Rev. B* **14**, 4964 (**1976**).

233 C. Bervillier and C. Godreche, *Phys. Rev. B* **21**, 5427 (**1980**).

234 D. Stauffer, M. Ferer, and M. Wortis, *Phys. Rev. Lett.* **29**, 345 (**1972**).

235 D. Beysens and A. Bourgou, *Universality of critical amplitude combinations and of correction-to-scaling ratios in binary fluids*. In: *Proceedings of the 8th Symposium on Thermophysical Properties* (Gaithersburg, Maryland, **1981**, pp. 383–388).

236 H. Guttinger and D. Cannell, *Phys. Rev. A* **24**, 3188 (**1981**).

237 D. Yu. Ivanov, *Dokl. Phys.* **53**, 241 (**2008**) [Dokl. Akad. Nauk **420**, 33 (**2008**)].

238 V. P. Skripov and M. Z. Faizullin, *Crystal–Liquid–Gas Phase Transitions and Thermodynamic Similarity* (Fizmatlit, Moscow, **2003**; Wiley-VCH, Berlin, Weinheim, **2006**).

239 M. Funke, R. Kleinrahm, and W. Wagner, *J. Chem. Thermodyn.* **34**, 735 (**2002**).

240 D. Yu. Ivanov: *Coexistence curve of quantum liquids*. In: *Abstracts of the XVIth International Conference on Chemical Thermodynamics* (Suzdal, Moscow, **2007**, vol. 1, pp. 92–93)).

241 V. P. Skripov, *Metastable Liquids* (Nauka, Moscow, **1972**; Wiley, New York, **1974**).

242 V. P. Skripov and A. V. Skripov, *Uspekhi Fiz. Nauk* **128**, 193 (**1979**).

243 V. P. Skripov, E. N. Sinitsyn, P. A. Pavlov, G. V. Ermakov, G. N. Muratov, N. V. Bulanov, and V. G. Baidakov, *Thermophysical Properties of Liquids in the Metastable State* (Atomizdat, Moscow, **1980**; Gordon and Breach, London, **1988**).

244 J. Osman and C. M. Sorensen, *J. Chem. Phys.* **73**, 4142 (**1980**).

245 D. Yu. Ivanov and V. A. Pavlov, *Teplofiz. Vysokikh Temperatur* **25**, 1014 (**1987**).

246 N. C. Ford, Jr. and G. B. Benedek: *The spectrum of light inelastically scattered by a fluid near its critical points*. In: *Proceedings of the Conference on Phenomena in the Neighborhood of Critical Points* (Edited by M. S. Green and J. V. Sengers, NBS Misc. Publ. 273, Washington **1966**, pp. 150–156); *Phys. Rev. Lett.* **15**, 649 (**1965**).

247 G. B. Benedek, *Optical mixing spectroscopy with applications to problems in physics, chemistry, biology and engineering*. In: *Polarisation, Matière et Rayonnement. Volume jubilaire en l'honneur d'Alfred Kastler* (Presses Universitaires de France, Paris, **1969**, pp. 49–84).

248 B. Chu, F. J. Schoenes, and M. E. Fisher, *Phys. Rev.* **185**, 219 (**1969**).

249 J. Kojima, N. Kuwahara, and M. Kaneko, *J. Chem. Phys.* **63**, 333 (**1975**).

250 K. Hamano, T. Kawazura, T. Koyama, and N. Kuwahara, *J. Chem. Phys.* **82**, 2718 (**1985**).

251 Y. Izumi and Y. Mijake, *Phys. Rev. A* **16**, 2120 (**1977**).

252 Y. Izumi, H. Sawano, and Y. Mijake, *Phys. Rev. A* **29**, 826 (**1984**).

253 P. Guenoun, R. Gastaud, F. Perrot, and D. Beysens, *Phys. Rev. A* **36**, 4876 (**1987**).

254 F. M. Kuni, *Statistical Physics and Thermodynamics* (Nauka, Moscow, **1981**).

255 I. R. Krichevsky, *Notions and Fundamentals of Thermodynamics* (Khimia, Moscow, **1970**).

256 J. W. Gibbs, *Thermodynamics. Statistical Mechanics* (Nauka, Moscow, **1982**).

257 R. Kubo, *Thermodynamics* (North-Holland, Amsterdam, **1968**; Mir, Moscow, **1970**).

258 D. Ter Haar and H. Wergeland, *Elements of Thermodynamics* (Addison-Wesley, Reading, MA, **1966**; Mir, Moscow, **1968**).

259 *Problems in Thermodynamics and Statistical Physics* (Edited by P. T. Landsberg, PION, London, **1971**; Mir, Moscow, **1974**).

260 H.-O. Peitgen and P. H. Richter, *The Beauty of Fractals* (Springer, Berlin, **1986**; Mir, Moscow, **1993**).

261 A. Yu. Grosberg and A. R. Khohlov, *Physics in the World of Polymers* (Nauka, Moscow, **1970**).

262 V. A. Rykov, *Zhurn. Fiz. Khimii* **59**, 2905 (**1985**).

263 D. Yu. Ivanov and V. A. Rykov, *Description of metastable states of matter in the critical region*. In: *Thermophysics of Metastable Liquids in Connection with Boiling and Crystallization Phenomena: Proceedings of the All-Union Symposium on Thermophysics of Metastable Liquids* (Sverdlovsk, Russia, **1985**, pp. 21–22).

264 A. M. Rubshteyn, V. G. Baidakov, and V. P. Skripov, *Isochoric heat capacity of liquid xenon near the critical point liquid–vapour, ibid*, pp. 33–34.

265 P. P. Bezverkhy, V. G. Martynets, E. V. Matizen, and V. F. Kukarin, *About the possibility of* ^{4}*He heat capacity description at the non-critical isochores by scaling dependences using spinodal curves, ibid*, pp. 19–20.

266 P. P. Bezverkhy, V. G. Martynets, E. V. Matizen, and V. F. Kukarin, *Zhurn. Eksper. Teor. Fiz.* **90**, 946 (**1986**).

267 L. Onsager, *Phys. Rev.* **37**, 405; **38**, 2265 (**1931**).

268 S. Chapman and T. G. Cowling, *The Mathematical Theory of Non-Uniform Gases* (Cambridge University Press, Cambridge, **1970**, 3rd edition; Gosinlitizdat, Moscow, **1960**).

269 J. O. Hirschfelder, C. F. Curtiss, and R. B. Bird, *Molecular Theory of Gases and Liquids* (Wiley, New York, **1954**; Gosinlitizdat, Moscow, **1961**).

270 J. V. Sengers, *Int. J. Thermophys.* **6**, 203 (**1985**).

271 P. Calmettes, Thèse de l'Université de Paris VI. Paris (**1978**).

272 P. Calmettes, *Phys. Rev. Lett.* **39**, 1151 (**1977**).

273 P. Calmettes, *Shear viscosity of fluids near a critical point*. In: *Proceedings of the 8th Symposium on Thermophysical Properties* (Gaithersburg, Maryland, **1981**, pp. 429–433).

274 L. H. Cohen, M. L. Dingus, and H. Meyer, *Phys. Rev. Lett.* **50**, 1058 (**1983**).

275 H. Meyer and L. H. Cohen, *Phys. Rev. A* **38**, 2081 (**1988**).

276 D. G. Friend and H. M. Roder, *Phys. Rev. A* **32**, 1941 (**1985**).

277 R. Mostert and J. V. Sengers, *Fluid Phase Equilib.* **75**, 235 (**1992**).

278 R. Folk and G. Moser, *J. Low Temp. Phys.* **99**, 11 (**1995**).

279 R. Folk and G. Moser, *Int. J. Thermophys.* **19**, 1003 (**1998**).

280 R. F. Berg and K. Gruner, *J. Chem. Phys.* **101**, 1513 (**1994**).

281 R. F. Berg and M. R. Moldover, *J. Chem. Phys.* **93**, 1926 (**1990**).

282 H. L. Swinney, D. L. Henry, and H. Z. Cummins, *J. de Physique* **33**, 81 (**1972**).

283 P. C. Hohenberg and B. I. Halperin, *Rev. Mod. Phys.* **49**, 435 (**1977**).

284 J. W. Rayleigh, *Phil. Mag.* **47**, 375 (**1899**).

285 M. F. Vuks, *Light Scattering in Gases, Liquids and Solutions* (Leningrad State University Publishing House, Leningrad, **1977**).

286 L. I. Mandelstam, *About optically homogeneous and turbid media*. In: *Collected Works* (Akad. Nauk SSSR, Leningrad, **1948**, vol. 1, pp. 109–124).

287 A. Einstein, *Ann. d. Physik. (Leipzig)* **33**, 1275 (**1910**).

288 M. V. Volkenshtein, *Molecular Optics* (Gostekhteorizdat, Moscow, Leningrad, **1951**).

289 B. Chu, *Ber. Bunsenges. Phys. Chem.* **76**, 202 (**1972**).

290 *Physical Encyclopaedia*. (Grand Russian Encyclopaedia, Moscow, **1988**, vol. 1, p. 704).

291 *Physical Encyclopaedia*. (Grand Russian Encyclopaedia, Moscow, **1998**, vol. 5, p. 760).

292 L. A. Zubkov and V. P. Romanov, *Uspekhi Fiz. Nauk* **154**, 615 (**1988**).

293 L. Landau and G. Placzek, *Phys. Zs. Sowjetunion* **5**, 172 (**1934**).

294 E. Gross, *Nature* **126**, 201 (**1930**).

295 *Solid State Optics and Ultrasound Physics: Collected Papers* (Edited by B. V. Novikov, St. Petersburg State University Publishing House, St. Petersburg, **1999**).

296 I. L. Fabelinsky, *Uspekhi Fiz. Nauk* **170**, 93 (**2000**).

297 B. V. Novikov, *Uspekhi Fiz. Nauk* **170**, 108 (**2000**).

298 P. Debye, *Phys. Rev. Lett.* **14**, 783 (**1965**).

299 P. C. Martin, *Many-Body Physics* (Gordon and Breach, New York, **1968**).

300 L. Mistura, *Nuovo Cim. B* **12**, 35 (**1972**).

301 W. Botch and M. Fixman, *J. Chem. Phys.* **42**, 199 (**1965**).

302 R. A. Ferrell, *Phys. Rev. Lett.* **24**, 1169 (**1970**).

303 R. Perl and R. A. Ferrell, *Phys. Rev. Lett.* **29**, 51 (**1972**).

304 B. J. Ackerson, C. M. Sorensen, R. C. Mockler, and W. J. O'Sullivan, *Phys. Rev. Lett.* **34**, 1371 (**1975**).

305 C. M. Sorensen, B. J. Ackerson, R. C. Mockler, and W. J. O'Sullivan, *Phys. Rev. A.* **13**, 1593 (**1976**).

306 K. Kawasaki, *Phys. Rev.* **150**, 291 (**1966**).

307 K. Kawasaki, *Ann. Phys. NY* **61**, 1 (**1970**).

308 K. Kawasaki, *Mode coupling and critical dynamics.* In:*Phase Transitions and Critical Phenomena* (Edited by C. Domb and M. S. Green, Academic Press, London, New York, **1976**, vol. 5a, pp. 165–403).

309 L. P. Kadanoff and J. Swift, *Phys. Rev.* **166**, 89 (**1968**).

310 *Physical Encyclopaedia.* (Grand Russian Encyclopaedia, Moscow, **1992**, vol. 2, p. 703).

311 K. Kawasaki, *Dynamical theory of fluctuations near critical points.* In: *Quantum Field Theory and Phase Transitions Physics* (Mir, Moscow, **1975**, pp. 101–148).

312 L. D. Landau and E. M. Lifshitz, *Hydrodynamics* (Nauka, Moscow, **1986**).

313 M. Fixman, *J. Chem. Phys.* **33**, 1357 (**1960**).

314 P. Bergé, P. Calmettes, C. Laj, and B. Voloshine, *Phys. Rev. Lett.* **23**, 693 (**1969**).

315 P. Bergé, P. Calmettes, C. Laj, M. Tournarie, and B. Voloshine, *Phys. Rev. Lett.* **24**, 1223 (**1970**).

316 P. Bergé and M. Dubois, *Phys. Rev. Lett.* **27**, 1125 (**1971**).

317 K. Kawasaki and S. M. Lo, *Phys. Rev. Lett.* **29**, 48 (**1972**).

318 D. W. Oxtoby and W. Gelbart, *J. Chem. Phys.* **61**, 2957 (**1974**).

319 *Photon Correlation and Light Beating Spectroscopy* (Edited by H. Z. Cummins and E. R. Pike, Plenum Press, New York, **1974**; Mir, Moscow, **1978**).

320 R. F. Chang, H. C. Burstyn, J. V. Sengers, and C. O. Alley, *Phys. Rev. Lett.* **27**, 1706 (**1971**).

321 J. K. Bhattacharjee and R. A. Ferrell, *Phys. Rev. A* **27**, 1544 (**1983**).

322 E. D. Siggia, B. I. Halperin, and P. C. Hohenberg, *Phys. Rev. B* **13**, 2110 (**1976**).

323 C. Paladin and L. Peliti, *J. de Physique Lett.* **43**, 15 (**1982**).

324 D. Beysens, A. Bourgou, and G. Paladin, *Phys. Rev. A* **30**, 2686 (**1984**).

325 D. L. Henry, L. E. Evans, and R. Kobayashi, *J. Chem. Phys.* **66**, 1802 (**1977**).

326 H. C. Burstyn, J. V. Sengers J. K. Bhattacharjee, and R. A. Ferrell, *Phys. Rev. A* **28**, 1567 (**1983**).

327 R. S. Basu and J. V. Sengers: *Viscosity of fluids near the gas–liquid critical point.* In: *Proceedings of the 8th Symposium on Thermophysical Properties* (Gaithersburg, Maryland, **1981**, pp. 434–439).

328 H. C. Burstyn, J. V. Sengers, and P. Esfandiari, *Phys. Rev. A* **22**, 282 (**1980**).

329 S.-H. Chen, J. Rouch, and P. Tartaglia: *Analysis of light scattering data on a binary liquid mixture of n-Hexane and Nitrobenzene in the critical region.* In: *Proceedings of the 8th Symposium on Thermophysical Properties* (Gaithersburg, Maryland, **1981**, pp. 449–450).

330 D. Beysens, G. Paladin, and A. Bourgou, *J. de Physique Lett.* **44**, 649 (**1983**).

331 M. E. Fisher, *Lecture notes presented at the "Advanced Course on Critical Phenomena"* at Merensky Institute, University of Stellenbosch, South Africa, **1982**.

332 C. A. Tracy and B. M. McCoy, *Phys. Rev. B* **12**, 368 (**1975**).

333 D. T. Jacobs, S. M. Y. Lau, A. Mukherjee, and C. A. Williams, *Int. J. Thermophys.* **20**, 877 (**1999**).

334 R. F. Chang, H. C. Burstyn, and J. V. Sengers, *Phys. Rev. A* **19**, 866 (**1979**).

335 V. P. Romanov and T. Kh. Salihov, *Optika i Spektroskopiya* **58**, 1091 (**1985**).

336 I. L. Fabelinsky, *Molecular Light Scattering* (Nauka, Moscow, **1965**).

337 E. L. Lakoza and A. V. Chalyi, *Zhurn. Eksper. Teor. Fiz.* **3**, 875 (**1977**).

338 V. L. Kuz'min, *Optika i Spektroskopiya* **39**, 546 (**1975**).

339 V. L. Kuz'min, *Optika i Spektroskopiya* **40**, 552 (**1976**).

340 V. L. Kuz'min, *Optika i Spektroskopiya* **44**, 529 (**1978**).

341 H. M. J. Boots, D. Bedeaux, and P. Mazur, *Physica A* **79**, 397 (**1975**).

342 H. M. J. Boots, D. Bedeaux, and P. Mazur, *Physica A* **84**, 217 (**1976**).

343 Yu. N. Barabanenkov and E. G. Stainova, *Optika i Spektroskopiya* **63**, 362 (**1987**).

344 Yu. N. Barabanenkov and E. G. Stainova, *Optika i Spektroskopiya* **59**, 1342 (**1985**).

345 Yu. N. Barabanenkov and E. G. Stainova, *Zhurn. Eksper. Teor. Fiz.* **88**, 1967 (**1985**).

346 V. L. Kuz'min and V. P. Romanov, *Optika i Spektroskopiya* **69**, 656 (**1990**).

347 L. L. Foldy, *Phys. Rev.* **67**, 107 (**1945**).

348 V. Twersky, *J. Math. Phys.* **3**, 700 (**1962**).

349 Yu. N. Barabanenkov, *Uspekhi Fiz. Nauk.* **117**, 49 (**1975**).

350 E. L. Lakoza and A. V. Chalyi, *Uspekhi Fiz. Nauk.* **140**, 393 (**1983**) [*Sov. Phys. Usp.* **26**, 573 (**1983**)].

351 V. L. Kuz'min, *Phys. Rep.* **123**, 365 (**1985**).

352 V. L. Kuz'min, V. P. Romanov, and L. A. Zubkov, *Phys. Rep.* **248**, 71 (**1994**).

353 K. S. Shifrin, *Light Scattering in a Turbid Medium* (Gostekhteorizdat, Moscow, Leningrad, **1951**).

354 L. A. Apresyan and Yu. A. Kravtsov, *Radiation-Transport Theory: Statistical and Wave Aspects* (Nauka, Moscow, **1983**).

355 A. Ishimaru, *Wave Propagation and Scattering in Random Media*, vol. **2**, *Multiple Scattering* (Academic Press, New York, **1978**; Mir, Moscow, **1981**).

356 V. I. Tatarsky, *Wave Propagation in a Turbulent Atmosphere* (Nauka, Moscow, **1967**).

357 R. B. Kopelman, R. W. Gammon, and M. R. Moldover, *Phys. Rev. A* **29**, 2048 (**1984**).

358 S. A. Casalnuovo, R. C. Mockler, and W. J. O'Sullivan, *Phys. Rev. A* **29**, 257 (**1984**).

359 H. C. Burstyn and J. V. Sengers, *Phys. Rev. A* **25**, 448 (**1982**).

360 D. Beaglehole, *Phys. Lett. A* **91**, 237 (**1982**).

361 S. B. Dierker and P. Wiltzius, *Phys. Rev. Lett.* **66**, 1185 (**1991**).

362 J. L. Han and M. Ciftan, *Phys. Lett. A* **124**, 495 (**1987**).

363 A. O. Parry and P. Evans, *Phys. Rev. Lett.* **64**, 439 (**1990**).

364 J. G. Shanks and J. V. Sengers, *Phys. Rev. A* **38**, 885 (**1988**).

365 W. I. Goldburg, *Light scattering study in liquids in the critical region*. In: *Light Scattering Near Phase Transitions* (Edited by H. Z. Cummins and A. P. Levanyuk, North-Holland, Amsterdam, **1983**; Nauka, Moscow, **1990**, pp. 337–375).

366 A. V. Chalyi, *Ukr. Fiz. Zhurn.* **13**, 1159 (**1968**).

367 D. W. Oxtoby and W. M. Gelbart, *J. Chem. Phys.* **60**, 3359 (**1974**).

368 D. W. Oxtoby and W. M. Gelbart, *Phys. Rev. A* **10**, 738 (**1974**).

369 D. Beysens and G. Zalczer, *Phys. Rev. A* **15**, 765 (**1977**).

370 L. A. Reith and H. L. Swinney, *Phys. Rev.* **12**, 1094 (**1975**).

371 A. J. Bray, *Phys. Rev. B* **14**, 1248 (**1976**).

372 A. J. Bray and R. F. Chang, *Phys. Rev. A* **12**, 2594 (**1975**).

373 L. V. Adzhemyan, L. Ts. Adzhemyan, L. A. Zubkov, and V. P. Romanov, *Pis'ma v Zhurn. Eksper. Teor. Fiz.* **22**, 11 (**1975**).

374 L. V. Adzhemyan, L. Ts. Adzhemyan, L. A. Zubkov, and V. P. Romanov, *Optika i Spektroskopiya* **46**, 967 (**1979**).

375 L. V. Adzhemyan, L. Ts. Adzhemyan, L. A. Zubkov, and V. P. Romanov, *Zhurn. Eksper. Teor. Fiz.* **78**, 1051 (**1980**).

376 L. V. Adzhemyan, L. Ts. Adzhemyan, L. A. Zubkov, and V. P. Romanov, *Zhurn. Eksper. Teor. Fiz.* **80**, 551 (**1981**).

377 D. Beysens and G. Zalczer, *Opt. Commun.* **26**, 172 (**1978**).

378 D. Beysens, Y. Garrabos, and F. Perrot, *Phys. Lett. A* **66**, 49 (**1978**).

379 R. C. Colby, L. M. Narducci, V. Bluemel, and J. Baer, *Phys. Rev. A* **12**, 1530 (**1975**).

380 L. Ts. Adzhemyan, V. P. Romanov, and T. Kh. Salihov, *Optika i Spektroskopiya* **58**, 339–345 (**1985**).

381 V. P. Romanov and T. Kh. Salihov, *Optika i Spektroskopiya* **59**, 1048, 1051–(**1985**).

382 C. M. Sorensen, R. C. Mockler, and W. J. O'Sullivan, *Phys. Rev. Lett.* **40**, 777 (**1978**).

383 C. M. Sorensen, R. C. Mockler, and W. J. O'Sullivan, *Phys. Rev. A* **17**, 2030 (**1978**).

384 A. Bøe and O. Lohne, *Phys. Rev. A* **17**, 2023 (**1978**).

385 A. Bøe and T. Sikkeland, *Phys. Rev. A* **16**, 2105 (**1977**).

386 D. Yu. Ivanov and A. F. Kostko, *Opt. Spectrosc. (USSR)* **55**, 573 (**1983**) [Optika i Spektroskopiya **55**, 950 (1983)].

387 D. Yu. Ivanov, A. F. Kostko, and V. A. Pavlov, *Sov. Phys. Dokl.* **30**, 397 (**1985**) [Dokl. Akad. Nauk SSSR **282**, 568 (1985)].

388 D. Yu. Ivanov, A. F. Kostko, and V. A. Pavlov: *Limiting case of scattering multiplicity in critical opalescence spectra.* In: *Liquid State Physics* (Kiev University Publishing House, Kiev, **1977**, N 14, pp. 121–128).

389 D. Yu. Ivanov, A. F. Kostko, and V. A. Pavlov: *Multiple scattering spectrum half-width calculation near a critical point* (VINITI, N 484, **1985**).

390 D. Yu. Ivanov, A. F. Kostko, and V. A. Pavlov: *Quasi-elastic multiple light scattering in water latex suspensions.* In: *Molecular and Biological Physics of Water Systems* (Leningrad State University Publishing House, Leningrad, **1986**, N 6, pp. 145–152).

391 A. F. Kostko, Ph.D. Dissertation (Leningrad State University, Leningrad, **1987**).

392 D. Yu. Ivanov, A. F. Kostko, and V. A. Pavlov, *Phys. Lett. A* **138**, 339 (**1989**).

393 D. Yu. Ivanov, *Critical opalescence: models, experiment.* In: *Proceedings International Conference on Mathematical Modelling* (Moscow, **2000**, vol. 1, pp. 48–63).

394 D. Yu. Ivanov, *Critical opalescence: Models-Experiment.* In: *Mathematical Modelling: Problems, Methods, Applications* (Kluwer Academic/Plenum Publishers, New York, **2001**, pp. 37–51).

395 D. Yu. Ivanov and A. V. Soloviev, *Laser radiation influence on the temperature of extra turbid media* (VINITI, N 1357-B, **1992**).

396 D. Yu. Ivanov, *Critical Opalescence: Application of the diffusion equation to multiple light scattering.* In: *Nucleation Theory and Applications* (Edited by J. W. P. Schmelzer, G. Röpke, and V. B. Priezzhev, Joint Institute for Nuclear Research, Publishing Department, Dubna, Russia, **2005**, pp. 295–407).

397 D. Yu. Ivanov, A. F. Kostko, and S. S. Proshkin, In: *Proceedings of the 13th European Conference on Thermophysical Properties.* Lisboa, Portugal 377 (**1993**).

398 D. Yu. Ivanov, A. F. Kostko, and S. S. Proshkin, *Critical opalescence spectrum in anilin-cyclohexane* (VINITI, N 24-B, **1992**).

399 B. J. Bern and R. Pecora, *Dynamic Light Scattering, with Applications to Chemistry, Biology, and Physics* (Wiley, New York, **1976**).

400 R. Pecora, *Dynamic Light Scattering: Applications of Photon Correlation Spectroscopy* (Plenum, New York, **1985**).

401 B. J. Bern and R. Pecora, *Dynamic Light Scattering* (Krieger, Malabar, FL, **1990**).

402 B. Chu, *Laser Light Scattering: Basic Principles and Practice* (Academic, San Diego, CA, **1991**).

403 *Dynamic Light Scattering: The Methods and Some Applications* (Edited by W. Brown, Clarendon, Oxford University Press, UK, **1993**).

404 G. S. Gorelik, *Dokl. Akad. Nauk SSSR* **58**, 45 (**1947**).

405 A. T. Forrester, W. E. Parkins, and E. Gerjuoy, *Phys. Rev.* **72**, 728 (**1947**).

406 E. Jakeman, C. J. Oliver, and J. M. Vaughan, *Opt. Commun.* **17**, 305 (**1976**).

407 D. Yu. Ivanov and A. F. Kostko, *Light scattering study in ternary butanol solutions.* In: *Molecular and Biological Physics of Water Systems* (Leningrad State University Publishing House, Leningrad, **1983**, N 5, pp. 51–56).

408 I. K. Yudin, G. L. Nikolaenko, V. I. Kosov, V. A. Agayan, M. A. Anisimov, and J. V. Sengers, *Int. J. Thermophys.* **18**, 1237 (**1997**).

409 A. M. Evtyushenkov, Yu. F. Kiyachenko, G. I. Olefirenko, and I. K. Yudin, *Pribory i Tehnika Eksperimenta* **5**, 157 (**1981**).

410 S. S. Vetohin, I. R. Gulakov, A. N. Percev, and I. V. Reznikov, *Single-Electron Photodetectors* (Atomizdat, Moscow, **1979**).

411 B. Crosignani, P. di Porto, and M. Bertolotti, *Statistical Properties of Scattered Light* (Academic Press, New York, **1975**).

412 E. Jakeman, *Photon correlations.* In: *Photon Correlation and Light Beating Spectroscopy* (Edited by H. Z. Cummins and E. R. Pike, Plenum Press, New York, **1974**; Mir, Moscow, **1978**).

413 E. B. Aleksandrov, Yu. M. Golubev, A. V. Lomakin, and V. A. Noskin, *Uspekhi Fiz. Nauk* **140**, 547 (**1983**).

414 P.-A. Lemieux and D. J. Durian, *Appl. Opt.* **40**, 3984 (**2001**).

415 R. G. W. Brown, *Appl. Opt.* **40**, 4004 (**2001**).

416 A. Lomakin, *Appl. Opt.* **40**, 4079 (**2001**).

417 B. Kruppa, *High Temp. – High Press.* **27/28**, 227 (**1995/1996**).

418 A. Leipertz, *Int. J. Thermophys.* **9**, 897 (**1988**).

419 A. Leipertz, *Fluid Phase Equlib.* **125**, 219 (**1996**).

420 H. L. Swinney and D. L. Henry, *Phys. Rev. A* **8**, 2586 (**1973**).

421 T. G. Braginskaya, V. V. Klyubin, V. A. Noskin, and N. M. Reinov, *Zhurn. Tehnich. Phys.* **50**, 785 (**1980**).

422 D. E. Koppel, *J. Chem. Phys.* **57**, 4814 (**1972**).

423 B. J. Frisken, *Appl. Opt.* **40**, 4087 (**2001**).

424 G. S. Getner and G. V. Flinn, *Pribory dlya nauchnyh issledovanii* **5**, 89 (**1975**).

425 V. V. Klyubin, V. A. Noskin, N. A. Saharova, and O. S. Chechik, *Zhurn. Tehnich. Phys.* **50**, 2433 (**1980**).

426 O. S. Chechik and V. V. Klyubin, *An Attempt of Latex Particles Diameter Measuring* (Izdatelstvo of "Znanie" Association, Leningrad, **1986**).

427 A. P. Ivanov, A. Ja. Khairullina, and A. P. Chaikovskaya, *Optika i Spektroskopiya* **35**, 1153 (**1973**).

428 C. M. Sorensen, R. C. Mockler, and W. J. O'Sullivan, *Phys. Rev. A* **14**, 1520 (**1976**).

429 I. L. Zel'manovich and K. S. Shifrin, *Tables of Light-Scattering; Vol. III: Damping, Scattering and Pressure Radiation Coefficients* (Gidrometeoizdat, Leningrad, **1968**).

430 F. Ya. Sid'ko, V. A. Zaharova, and V. N. Lopatin, *Integral Indicatrix of Light-Scattering on "Soft" Spherical Particles* (Novosibirsk, Nauka, **1977**).

431 Yu. A. Bal'tsevich, V. G. Martynets, and E. V. Matizen, *Zhurn. Eksper. Teor. Fiz.* **51**, 983 (**1966**).

432 M. A. Anisimov, Yu. F. Kiyachenko, G. L. Nikolaenko, and I. K. Yudin, *Inzhenerno Fiz. Zhurn.* **38**, 652 (**1980**).

433 S. Will and A. Leipertz, *Int. J. Thermophys.* **16**, 433 (**1995**).

434 V. G. Martynets and E. V. Matizen, *Zhurn. Eksper. Teor. Fiz.* **58**, 430 (**1970**).

435 K. B. Lyons, R. C. Mockler, and W. J. O'Sullivan, *Phys. Rev. Lett.* **30**, 42 (**1973**).

436 D. Yu. Ivanov, A. F. Kostko, and V. A. Pavlov: *Study of multiple scattering spectra in water suspensions by photon correlation method.* In: *Optics of Sea and Atmosphere Proceedings of the IXth Plenum of the Working Group on Ocean Optics of the Academy of Sciences of USSR Commission on the World Ocean Problems* (Leningrad, **1984**, pp. 155–156).

437 D. Yu. Ivanov, A. F. Kostko, and V. A. Pavlov, *Diffusion approximation display in multiple scattering spectra of laser radiation.* In: *Thermophysical Properties of Substances: Proceedings of the 3rd All-Union Symposium on Propagation of Laser Radiation in Disperse Media* (Novosibirsk, **1985**, vol. 2, pp. 71–74).

438 D. Yu. Ivanov, A. F. Kostko, and V. A. Pavlov: *Study of turbid disperse systems by the method of correlation spectroscopy.* In: *Proceedings of the XIVth All-Union Conference on Radio Wave Propagation* (Nauka, Leningrad **1984**, vol. 2, pp. 126–128).

439 C. W. J. Beenakker and P. Mazur, *Physica A* **120**, 388 (**1983**).

440 V. A. Pavlov, *Optika i Spektroskopiya* **64**, 828 (**1988**).

441 G. Maret and P.-E. Wolf, *Z. Phys. B* **65**, 409 (**1987**).

442 D. Y. Pine, D. A. Weitz, P. M. Chaikin, and E. Herbolzheimer, *Phys. Rev. Lett.* **60**, 1134 (**1988**).

443 I. Freund, M. Kaveh, and N. Rosenbluh, *Phys. Rev. Lett.* **60**, 1130 (**1988**).

444 I. Edrei and M. Kaveh, *Phys. Rev. B* **38**, 950 (**1988**).

445 M. J. Stephen, *Phys. Lett. A* **127**, 371 (**1988**).

446 M. J. Stephen, *Phys. Rev. B* **37**, 1 (**1988**).

447 D. A. Weitz, D. Y. Pine, P. N. Pusey, and R. J. A. Tough, *Phys. Rev. Lett.* **63**, 1747 (**1989**).

448 F. C. MacKintosh and S. John, *Phys. Rev. B* **40**, 2383 (**1989**).

449 X. Qiu, X. L. Wu, D. Y. Pine, D. A. Weitz, and P. M. Chaikin, *Phys. Rev. Lett.* **65**, 516 (**1990**).

450 S. Fraden and G. Maret, *Phys. Rev. Lett.* **65**, 512 (**1990**).

451 D. G. Dalgleish and D. S. Horne, *Milchwissenschaft* **46**, 417 (**1991**).

452 D. A. Weitz and D. Y. Pine, In: *Dynamic Light Scattering: The Methods and some Applications* (Edited by W. Brown, Clarendon, Oxford University Press, UK, **1993**, pp. 652–720).

453 D. A. Weitz, J. X. Zhu, D. J. Durian, H. Guang, and D. Y. Pine, *Physica Scripta* **49**, 610 (**1993**).

454 P. D. Kaplan, M. H. Kao, A. G. Yodh, and D. J. Pine, *Appl. Opt.* **32**, 3828 (**1993**).

455 P. Štepánek, *J. Chem. Phys.* **99**, 6384 (**1993**).

456 J. C. Earnshaw and A. H. Jaafar, *Phys. Rev. E* **49**, 5408 (**1994**).

457 D. J. Durian, *Phys. Rev. E* **51**, 3350 (**1995**).

458 D. J. Durian, *Physica A* **229**, 218 (**1996**).

459 T. G. Mason, H. Guang, and D. A. Weitz, *J. Opt. Soc. Am. A* **14**, 139 (**1997**).

460 G. Maret, *Curr. Opin. Colloid Interface Sci.* **2**, 251 (**1997**).

461 P.-A. Lemieux, M. U. Vera, and D. J. Durian, *Phys. Rev. E* **57**, 4498 (**1998**).

462 G. Nissato, P. Hebraud, J.-P. Munch, and S. J. Candau, *Phys. Rev. E* **61**, 2879 (**2000**).

463 L. Vanel, P.-A. Lemieux, and D. J. Durian, *Appl. Opt.* **40**, 4179 (**2001**).

464 R. G. Newton, *Scattering Theory of Waves and Particles* (McGraw-Hill, New York, **1966**; Mir, Moscow, **1969**).

465 M. Planck, *Sitzungsber. Akad. Wiss. Berlin* **10**, 151 (**1896**).

466 M. Planck, *Sitzungsber. Akad. Wiss. Berlin* **22**, 740 (**1904**).

467 K. M. Case and P. F. Zweifel, *Linear Transport Theory* (Addison-Wesley, Reading, MA, **1967**; Mir, Moscow, **1972**).

468 A. Ishimaru, *Radio Sci.* **10**, 45 (**1975**).

469 S. Chandrasekhar, *Stochastic Problems in Physics and Astronomy* (*Rev. Mod. Phys.* **15**, 1 (**1943**); Gostekhteorizdat, Moscow, Leningrad, **1948**).

470 M. V. Volkenshtein, *Conformation Statistics of Polymer Chains* (Izdatelstvo Akademii Nauk SSSR, Moscow, **1959**).

471 J. M. Ziman, *Models of Disorder* (Cambridge University Press, Cambridge, London, NY, Melbourne, **1979**; Mir, Moscow, **1982**).

472 P.-G. de Gennes, *Scaling Concepts in Polymer Physics* (Cornell University Press, Ithaca and London, **1979**; Mir, Moscow, **1982**).

473 S. E. Bresler and B. L. Erusalimsky, *Physics and Chemistry of Macromolecules* (Nauka, Moscow, Leningrad, **1963**).

474 E. Fermi, *Nuclear Physics* (University Chicago Press, Chicago, **1950**; Gosinlitizdat, Moscow, **1951**).

475 G. I. Bell and S. Glasstone, *Nuclear Reactor Theory* (Reinhold, **1970**; Atomizdat, Moscow, **1974**).

476 S. B. Shihov, *Problems of Mathematical Theory of Reactors: Linear Analysis* (Atomizdat, Moscow, **1973**).

477 S. B. Shihov and V. B. Troyansky, *The Theory of Nuclear Reactors. Gas-Kinetic Theory* (Energoatomizdat, Moscow, **1983**).

478 V. A. Ambartsumyan, *Dokl. Akad. Nauk Arm SSR* **8**, 101 (**1948**).

479 D. Yu. Ivanov, A. F. Kostko, V. A. Ryzhov, and V. A. Pavlov, *Determination of mean dimension and concentration of particles of dispersed systems by the method of multiple scattering photon correlation spectroscopy*. In: *Proceedings of IVth All-Union Symposium on Propagation of Laser Radiation in Disperse Media* (Obninsk, Barnaul, **1988**, vol. 2, pp. 11–13).

480 D. Yu. Ivanov, A. F. Kostko, V. A. Pavlov, and V. A. Ryzhov, *Method of determination of mean dimension and concentration of suspended particles* (1390539 Patent of USSR, MKI⁴ G 01 N 15/02 publ. 23. 04. 88, Bull. N 15).

481 V. A. Pavlov, *Diffusion-transport equation replacement at multiple scattering spectra description*. In: *Proceedings of IVth All-Union Symposium on Laser Radiation Propagation in Disperse Medium* (Obninsk, Barnaul, **1988**, vol. 2, pp. 157–159).

482 P. M. Morse and H. Feshbach, *Methods of Theoretical Physics* (McGraw-Hill, New York, **1953**; Gosinlitizdat, Moscow, **1960**).

483 H. S. Carslaw and J. C. Jaeger, *Conduction of Heat in Solids*, 2nd edition, (Oxford University Press, USA, **1986**; Nauka, Moscow, **1974**).

484 *Applications of the Monte Carlo Method in Statistical Physics* (Edited by K. Binder, Springer, Berlin, Heidelberg, New York, Tokyo, **1984**; Mir, Moscow, **1982**).

485 A. F. Kostko and V. A. Pavlov, *Izvestiya Ross. Akad. Nauk, Seria Fiz.* **60**, 162 (**1996**).

486 A. F. Kostko and V. A. Pavlov, *Appl. Opt.* **36**, 7577 (**1997**).

487 N. N. Bogoliubov and D. V. Shirkov, *Introduction to the Theory of Quantized Fields* (Nauka, Moscow, **1984**, 4th edition).

488 M. E. Fisher and R. J. Burford, *Phys. Rev.* **156**, 583 (**1967**).

489 S. S. Alpert, Y. Yeh, and E. Lipworth, *Phys. Rev. Lett.* **14**, 486 (**1965**).

490 S. S. Alpert, *Time-dependent concentration fluctuations near the critical temperature*. In: *Proceedings of the Conference on Phenomena in the Neighborhood of Critical Points* (Edited by M. S. Green and J. V. Sengers, NBS Misc. Publ. 273, Washington **1966**, pp. 157–160).

491 M. A. Ayzerman, *Theory of Automatic Control* (Nauka, Moscow, **1967**).

492 U. Tietze and Ch. Schenk, *Halbleiter-Schaltungstechnik* (Springer, Berlin, **1980**; Mir, Moscow, **1982**).

493 G. L. Nikolaenko, Ph.D. Dissertation (Leningrad State University, **1987**).

494 L. A. Zubkov, V. P. Romanov, and T. Kh. Salihov, *Optika i Spektroskopiya* **68**, 110 (**1990**).

495 C. Bendjaballah, *Opt. Commun.* **9**, 279 (**1973**).

496 J. G. Rarity, *J. Chem. Phys.* **85**, 733 (**1986**).

497 J. K. G. Dhont and C. G. de Kruif, *J. Chem. Phys.* **79**, 1658 (**1983**).

498 H. C. Burstyn and J. V. Sengers, *Phys. Rev. A* **27**, 1071 (**1983**).

499 H. Brumberger, *Scattering of light and X-rays from critically opalescent systems.* In: *Proceedings of the Conference on Phenomena in the Neighborhood of Critical Points* (Edited by M. S. Green and J. V. Sengers, NBS Misc. Publ. 273, Washington, **1966**, pp. 116–123).

500 H. Z. Cummins, N. Knable, and Y. Yeh, *Phys. Rev. Lett.* **12**, 150 (**1964**).

501 J. V. Sengers, *Transport properties of fluids near critical points.* In: *Critical Phenomena, International School of Physics "Enrico Fermi"*, LI Course (Edited by M. S. Green, Academic Press, New York, **1971**, pp. 445–507).

502 J. V. Sengers and P. H. Keyes, *Phys. Rev. Lett.* **26**, 70 (**1971**).

503 G. Arcovito, C. Faloci, M. Roberti, and L. Mistura, *Phys. Rev. Lett.* **22**, 1040 (**1969**).

504 J. V. Sengers, *Behavior of viscosity and thermal conductivity of fluids near the critical point.* In: *Proceedings of the Conference on Phenomena in the Neighborhood of Critical Points* (Edited by M. S. Green and J. V. Sengers, NBS Misc. Publ. 273, Washington **1966**, pp. 165–177).

505 D. McIntyre and A. M. Wims, In: *Proceedings of Interdisciplinary Conference on Electromagnetic Scattering, II* (Edited by R. S. Stein, New York, **1967**, p. 457).

506 J. M. H. Levelt Sengers, *Solutions near the solvent's critical point: a summary.* In: *Fluides Supercritiques et Materiaux* (Edited par J.-P. Petitet et F. Cansell, Ecole d'Ete, Saint-Boil, France, **1995**, pp. 9–16).

507 R. I. Efremova and E. V. Matizen, *Zhurn. Eksper. Teor. Fiz.* **91**, 149 (**1986**).

508 E. V. Matizen, P. P. Bezverkhy, and V. G. Martynets, *Zhurn. Struct. Khimii* **39**, 655 (**1998**).

509 E. V. Matizen, P. P. Bezverkhy, and V. G. Martynets, *Phys. Rev. E* **59**, 2927 (**1999**).

510 L. A. Guildner, *Proc. Natl. Acad. Sci.* **44**, 1149 (**1958**).

511 R. Reid, D. Prausnitz, and T. Sherwood, *Properties of Gases and Liquids* (McGraw-Hill, New York, **1977**; Khimiya, Leningrad, **1982**).

512 A. Michels, J. V. Sengers, and P. S. van der Gulik, *Physica* **28**, 1201; 1216; 1238 (**1962**).

513 I. F. Golubev and V. P. Sokolova, *Teploenergetika* **11**, 64 (**1964**).

514 D. Needham and H. Ziebland, *Int. J. Heat Mass Transfer* **8**, 1387 (**1965**).

515 G. B. Benedek, *Thermal fluctuations and the scattering of light.* In: *Statistical Physics, Phase Transitions, and Superfluidity. Brandeis University Summer Institute in Theoretical Physics* (Edited by M. Chretien, S. Deser, and E. P. Gross, Gordon and Breach, New York, **1968**, vol. 2, p. 1).

516 C. A. Nieto de Castro and H. M. Roder: *Thermal conductivity of argon at 300.65 K. Evidence for a critical enhancement?* In: *Proceedings of the 8th Symposium on Thermophysical Properties* (Gaithersburg, Maryland, **1981**, pp. 241–246).

517 N. J. Trappeniers, *The behavior of the coefficient of heat conductivity in the critical region of xenon and argon.* In: *Proceedings of the 8th Symposium on Thermophysical Properties* (Gaithersburg, Maryland, **1981**, pp. 232–240).

518 R. Tufeu, D. Y. Ivanov, Y. Garrabos, and B. Le Neindre, *Ber. Bunsenges. Phys. Chem.* **88**, 422 (**1984**).

519 J. M. H. Levelt Sengers, *Critical behavior of fluids, concepts and applications.* In: *Supercritical Fluids: Fundamentals for Application, NATO ASI* (Edited by E. Kiran and J. M. H. Levelt Sengers, Kluwer, Dordrecht, Netherlands, **1994**, pp. 3–37).

520 L. P. Kadanoff and P. C. Martin, *Ann. Phys. NY* **24**, 419 (**1963**).

521 C. A. Croxton, *Liquid State Physics – A Statistical Mechanical Introduction* (Cambridge University Press, Cambridge, **1974**; Mir, Moscow, **1978**).

522 *Transport Properties of Fluids: Their Correlation, Prediction and Estimation* (Edited by J. Millat, J. H. Dymond, and A. Nieto de Castro, Cambridge University Press, Cambridge, **1996**).

523 N. A. Smirnova, *Molecular Theories of Solutions* (Khimia, Leningrad, **1987**).

524 Yu. K. Tovbin and M. M. Senyavin: *Cluster approach to the calculation of complex liquid systems phase diagram.* In: *Abstracts of the International Conference on Chemical Thermodynamics* (St. Petersburg, **2002**, p. 388).

525 A. Parola and L. Reatto, *Phys. Rev. Lett.* **53**, 2417 (**1984**).

526 A. Parola and L. Reatto, *Phys. Rev. A* **31**, 3309 (**1985**).

527 R. Folk and G. Moser, *Int. J. Thermophys.* **16**, 1363 (**1995**).

528 D. Pini, A. Parola, and L. Reatto, *Int. J. Thermophys.* **19**, 1545 (**1998**).

529 G. A. Olchowy and J. V. Sengers, *Phys. Rev. Lett.* **61**, 15 (**1988**).

530 G. A. Olchowy and J. V. Sengers, *Int. J. Thermophys.* **10**, 417 (**1989**).

531 J. V. Sengers, J. Luettmer-Strathmann: *Critical enhancements.* In: *Transport Properties of Fluids: Their Correlation, Prediction and Estimation* (Edited by J. Millat, J. H. Dymond, and C. A. Nieto de Castro, Cambridge University Press, Cambridge, **1996**, p. 113).

532 L. P. Filippov, *Methods of Computation and Prediction of the Properties of Matter* (Moscow State University Publishers, Moscow, **1988**).

533 B. J. Baily and K. Kellner, *Physica* **39**, 444 (**1968**).

534 *Experimental Thermodynamics, Vol. II. Measurement of the Transport Properties of Fluids* (Edited by W. A. Wakeham, A. Nagashima, and J. V. Sengers, Blackwell Scientific, Oxford, **1991**).

535 A. Nagashima, *Int. J. Thermophys.* **16**, 1069 (**1995**).

536 K. Kraft, S. Will, and A. Leipertz, *Measurement* **14**, 135 (**1994**).

537 K. Kraft, M. Matos Lopes, and A. Leipertz, *Int. J. Thermophys.* **16**, 423 (**1995**).

538 R. Tufeu and D. Ivanov, *Spectroscopie en milieux supercritique* (Edited par J.-P. Petitet et F. Cansell, Ecole d'Ete, Saint-Boil, France, **1995**, pp. 110–141).

539 M. Faraday, *Phil. Trans.* **135**, 170 (**1845**).

540 J. Dewar, *Phil. Mag.* **18**, 210 (**1884**).

541 P. Davies, *Ammonia.* In: *Thermodynamic Functions of Gases* (Butterworths Scientific, London, **1956**, vol. 1, pp. 33–87).

542 L. Haar and J. S. Gallagher, *J. Phys. Chem. Ref. Data* **7**, 635 (**1978**).

543 R. Mostert, H. R. van den Berg, and P. S. van der Gulik, *Rev. Sci. Instr.* **60**, 3466 (**1989**).

544 R. Mostert, H. R. van den Berg, and P. S. van der Gulik, *Int. J. Thermophys.* **10**, 409 (**1989**).

545 R. Mostert, H. R. van den Berg, P. S. van der Gulik, and J. V. Sengers, *J. Chem. Phys.* **92**, 5454 (**1990**).

546 S. H. Jawad, M. J. Dix, and W. A. Wakeham, *Int. J. Thermophys.* **20**, 45 (**1999**).

547 B. Le Neindre, Ph.D. thesis (University of Paris, Paris, **1969**).

548 R. Tufeu, B. Le Neindre, and P. Bury, *C.R. Acad. Sci. Paris* **272**, 113 (**1971**).

549 R. Tufeu, B. Le Neindre, and P. Bury, *C.R. Acad. Sci. Paris* **273**, 61 (**1971**).

550 B. Le Neindre, R. Tufeu, P. Bury, and J. V. Sengers, *Ber. Bunsenges. Phys. Chem.* **77**, 262 (**1973**).

551 B. Le Neindre, P. Bury, R. Tufeu, and B. Vodar, *J. Chem. Eng. Data* **21**, 265 (**1976**).

552 R. Tufeu and B. Le Neindre, *Int. J. Thermophys.* **1**, 375 (**1980**).

553 B. Le Neindre, Y. Garrabos, and R. Tufeu, *Ber. Bunsenges. Phys. Chem.* **88**, 916 (**1984**).

554 B. Le Neindre, Y. Garrabos, and R. Tufeu, *Int. J. Thermophys.* **12**, 307 (**1991**).

555 D. Vidal, R. Tufeu, Y. Garrabos, and B. Le Neindre, *Thermophysical properties of noble gases at room temperatures up to 1 GPa.* In: *High Pressure Science and Technology* (Edited by B. Vodar and Ph. Marteau, Pergamon, Oxford, UK, **1980**, pp. 692–701).

556 N. B. Vargaftik, L. P. Filippov, A. A. Tarzimanov, and E. E. Tocky, *Reference Book on Thermal Conductivity of Liquids and Gases* (Energoatomizdat, Moscow, **1990**).

557 E. Shmidt, *Int. J. Heat Mass Transfer* **1**, 92 (**1960**).

558 *The Theory of Heat Exchange. Terminology* (Izdatelstvo Standartov, Moscow, **1971**).

559 S. S. Kutateladze, *Similarity Analysis and Physical Modelling* (Nauka, Novosibirsk, **1986**).

560 R. Tufeu, A. Letaief, and B. Le Neindre, *Turbidity, thermal diffusivity and thermal conductivity of ammonia along the critical isochore.* In: *Proceedings of the 8th Symposium on Thermophysical Properties* (Gaithersburg, Maryland, **1981**, pp. 451–457).

561 M. F. Vuks, *Optika i Spektroskopiya* **25**, 857 (**1968**).

562 Y. Rocard, *Ann. Phys. (Paris)* **10**, 116 (**1928**).

563 Y. Rocard, *J. Phys. Radium* **4**, 165 (**1933**).

564 J. Yvon, *La propagation et la diffusion de la lumière* (Hermann, Paris, **1937**).

565 M. Giglio and A. Vendramini, *Phys. Rev. Lett.* **35**, 168 (**1975**).

566 D. Beysens, *J. Chem. Phys.* **64**, 2579 (**1976**).

567 C. Houessou, P. Guenoun, R. Gastaud, F. Perrot, and D. Beysens, *Phys. Rev. A* **32**, 1818 (**1985**).

568 H. J. M. Hanley, J. V. Sengers, and J. F. Ely, In: *Thermal Conductivity, 14th Proceedings* (Edited by P. G. Klemens and T. K. Chu, Plenum Press, New York, **1976**, p. 383).

569 C. S. Gragoe and D. R. Harper, *Sci. Pap. Bur. Stand.* **17**, 287 (**1921**).

570 H. Wiese and D. Horn, *J. Chem. Phys.* **94**, 6429 (**1991**).

571 E. R. van Keuren, H. Wiese, and D. Horn, *Ber. Bunsenges. Phys. Chem.* **98**, 269 (**1994**).

572 J. A. Lock, *Appl. Opt.* **40**, 4187 (**2001**).

573 G. Popescu and A. Dogariu, *Appl. Opt.* **40**, 4215 (**2001**).

574 A. Wax, C. Yang, R. R. Dasari, and M. S. Feld, *Appl. Opt.* **40**, 4222 (**2001**).

575 E. J. Nijman, H. G. Merkus, J. C. M. Marijnissen, and B. Scarlett, *Appl. Opt.* **40**, 4058 (**2001**).

576 D. J. Durian and J. Rudnick, *J. Opt. Soc. Am. A* **14**, 235 (**1997**).

577 A. A. Cox and D. J. Durian, *Appl. Opt.* **40**, 4228 (**2001**).

578 *Measurements of Suspended Particles by Quasi-Elastic Light Scattering* (Edited by B. E. Dahneke, Wiley, New York, **1984**).

579 *Modern Methods of Particle Size Analysis* (Edited by H. G. Barth, Wiley-Interscience, New York, **1984**).

580 *Optical Particle Sizing* (Edited by G. Gouesbet, Plenum, **1988**).

581 *Non-destructive Testing* (Edited by J. M. Farley and R. W. Nichols, Pergamon, Oxford, **1988**).

582 *An Introduction to Dynamic Light Scattering by Macromolecules* (Academic, San Diego, **1990**).

583 T. Allen, *Particles Size Measurement*, 4th edition, (Chapman and Hall, New York, **1990**).

584 D. Yu. Ivanov, A. F. Kostko, V. A. Pavlov, and S. S. Proshkin, *Complex optical method of study of processes in colloids.* In: *Abstracts of "Optics, Glass, Laser" Conference* (St. Petersburg, **1995**, pp. 94–95).

585 A. F. Kostko, D. Y. Ivanov, V. A. Pavlov, and S. S. Proshkin, *Diffusing photon correlation monitoring of particle motion in evaporated latex.* In: *Abstracts of the 71st Colloid and Surface Science Symposium* (Newark, Delaware, **1997**, p. 73).

586 A. V. Lebedev, *Colloidal Chemistry of Synthetic Latexes* (Khimia, Leningrad, **1987**).

587 M. A. McHugh and V. J. Crukonis, *Supercritical Fluid Extraction: Principles and Practice*, 2nd edition (Butterworth, Stoneham, MA, **1994**).

588 V. P. Saraf and E. Kiran, *J. Supercrit. Fluids*, 37 (**1988**).

589 Y. Garrabos, B. Le Neindre, P. Subra, F. Cansell, and C. Pommier, *Ann. Chim. France* **17**, 55 (**1992**).

590 T. Koga, S. Zhou, and B. Chu, *Appl. Opt.* **40**, 4170 (**2001**).

591 D. Yu. Ivanov and R. Tufeu, In: *Proceedings of the 3rd International Symposium on Supercritical Fluids*, Strasbourg, France, 327 (**1994**).

592 D. Yu. Ivanov, R. Tufeu, and A. V. Soloviev: *Particle formation in polymer–supercritical fluid systems.* In:

Proceedings of the 14th European Conference on Thermophysical Properties (Lyon, Villeurbanne, France, **1996**, p. 190).

593 D. Yu. Ivanov, R. Tufeu, and A. V. Soloviev: *Static and dynamic light scattering test of polymer particle formation process in supercritical fluids.* In: *Process Technology Proceedings* (Edited by Ph. Rudolf von Rohr and Ch. Trepp, Elsevier, The Netherlands, **1996**, vol. 12, pp. 389–391).

594 I. K. Yudin, G. L. Nikolaenko, E. E. Gorodetskii, V. R. Melikyan, E. L. Markhashov, V. A. Agayan, M. A. Anisimov, and J. V. Sengers, *Physica A* **251**, 235 (**1998**).

595 E. G. Burya, I. K. Yudin, V. A. Dechabo, V. I. Kosov, and M.

A Anisimov, *Appl. Opt.* **40**, 4028 (**2001**).

596 A. Ja. Khairullina, A. N. Korolevich, and T. V. Oleinik, *Proc. SPIE* **1981**, 38 (**1992**).

597 A. Ja. Khairullina, A. N. Korolevich, and T. V. Oleinik, *Proc. SPIE* **1884**, 124 (**1993**).

598 M. Kallala, C. Sanchez, and B. Cabane, *Phys. Rev. E* **48**, 3692 (**1993**).

599 A. V. Soloviev, R. Tufeu, and D. Yu. Ivanov, *Dokl. Physical Chem.* **374**, 190 (**2000**) [Dokl. Akad. Nauk **374**, 510 (**2000**)].

600 A. V. Soloviev, Thèse de l'Université Paris 13. Paris, (**2000**).

601 A. Soloviev, D. Ivanov, R. Tufeu, and A. V. Kanaev, *J. Mater. Sci. Lett.* **20**, 905 (**2000**).

Index

Critical Behavior of Nonideal Systems. Dmitry Yu. Ivanov
Copyright © 2008 WILEY-VCH Verlag GmbH & Co. KGaA, Weinheim
ISBN: 978-3-527-40658-6